The Relevance of the Time Domain to Neural Network Models

Springer Series in Cognitive and Neural Systems

Volume 3

Series Editors
John G. Taylor
King's College, London, UK

Vassilis Cutsuridis
Boston University, Boston, MA, USA

For further volumes:
www.springer.com/series/8572

A. Ravishankar Rao • Guillermo A. Cecchi
Editors

The Relevance of the Time Domain to Neural Network Models

Springer

Editors
A. Ravishankar Rao
IBM Thomas J. Watson Research Center
1101 Kitchawan Road
Yorktown Heights, NY 10598,
USA
ravirao@us.ibm.com

Guillermo A. Cecchi
Dept. Silicon Technology
IBM Thomas J. Watson Research Center
1101 Kitchawan Road
Yorktown Heights, NY 10598,
USA
gcecchi@us.ibm.com

ISBN 978-1-4614-0723-2   e-ISBN 978-1-4614-0724-9
DOI 10.1007/978-1-4614-0724-9
Springer New York Dordrecht Heidelberg London

Library of Congress Control Number: 2011938345

© Springer Science+Business Media, LLC 2012
All rights reserved. This work may not be translated or copied in whole or in part without the written permission of the publisher (Springer Science+Business Media, LLC, 233 Spring Street, New York, NY 10013, USA), except for brief excerpts in connection with reviews or scholarly analysis. Use in connection with any form of information storage and retrieval, electronic adaptation, computer software, or by similar or dissimilar methodology now known or hereafter developed is forbidden.
The use in this publication of trade names, trademarks, service marks, and similar terms, even if they are not identified as such, is not to be taken as an expression of opinion as to whether or not they are subject to proprietary rights.

Printed on acid-free paper

Springer is part of Springer Science+Business Media (www.springer.com)

Foreword

What is the relevance of temporal signal structure to the brain? We may gain some insight by comparing the brain to the computer. In the modern computer, signals are binary (have only two possible values), are made to change as quickly as technology permits, and temporal relations between signals are of central importance. The computer is driven by a clock through a quick succession of globally ordered states, while great care and effort is expended to make sure that no signal spills over from one state to the next. Ordered states are defined by commands in a program, each command specifying the setting of a large number of switches. At one time [1], this picture of a digital machine was taken seriously as a model for the brain, switches being identified with neurons. Digital machines are universal, meaning that any conceivable finite process can be realized in them, thus creating the vision that also the processes of the mind could be realized as processes in a physical machine. At the time, this idea was taken as the breakdown of the formerly perceived impenetrable glass wall between mind and matter. Unfortunately, the research program of Artificial Intelligence, which was built on this vision, has not given us intelligence in the machine yet. What is wrong with this vision of the brain as a digital machine? The succession of states in the computer is specified by programs, programs arise in human brains, and thus processes in the computer are imposed on it from outside. The big remaining question regarding the brain is that of the origin of its ordered states and sequences of states.

The role of temporal signal correlations in the brain may well be compared to that in the computer. The purpose of the brain is to coordinate activity in its various parts into ordered states and successions of states, such that things that belong together and form part of a functional whole are activated together. In this task of coordination, the brain is essentially out on its own, with very scant external help, which can in no way be compared to the insight of the computer's programmer. Classical artificial neural network models (important examples being the perceptron and associative memory) tended to grossly underestimate this task of generating and organizing brain states. In these models, time is paced by the presentation of stimuli, the network responding to each input pattern by convergence to a stationary state. This volume concentrates on a different brand of neural network models, in which the generation of temporal patterns is the focus of interest. As these studies in their turn tend to pay less attention to the solution of functional tasks (beyond the standard problem of segmentation) and concentrate to a large extent on the modeling of brain rhythms that are actually found, it may be of interest if I attempt to give a wider perspective on the functional significance of temporal signal structure.

There are two aspects to the data structure of brain state, that is, to the way neural activity represents cognitive content. Considering neurons as elementary symbols, these aspects are (a) which of these symbols are active in a given psychological moment, and (b) how these symbols are put in relation to each other. If there are several objects in a scene, for example, each to be described by several attributes, a number of neurons will be active to represent the objects and the attributes (aspect (a)), but it is also necessary to represent the information which of the several attributes refer to which of the several objects (aspect (b)). Another example is visual (or more generally, sensory) segmentation: the problem of expressing the subdivision of the sensory field into coherent perceptual objects.

This is generally called the binding problem—the problem of representing relatedness between the symbols represented by neurons. It is now common lore to consider neural signal synchrony as solution to the binding problem: sets of neurons that are relating to each other express this by firing simultaneously. In simple cases, such as the above examples, this seems a perfect solution, as both generation and functional exploitation of signal synchrony are natural to neural networks. Signal synchrony is generated by plausibly existing neural connections. In object-attribute binding, the branching feed-forward connections from the original stimuli to neurons representing objects and attributes can propagate the same signal fluctuations to those neurons as signature of common origin and as expression of relations between attributes and objects, In sensory segmentation, horizontal connections between the neurons in a sensory field, being shaped by spatial closeness and other Gestalt laws, tend to run between neurons responding to the same perceptual object, and these connections thus tend to correlate signals within segments, as has been modelled many times. Functional exploitation, that is, the read-out of signal synchrony, relies on the fact that neurons are coincidence detectors, and thus functional interaction is restricted to sets of signals that are synchronous.

As nice and conceptually coherent the picture engendered by these examples is, it doesn't settle the binding issue, for experimental and for theoretical reasons. It is a disturbing fact that in spite of intensive search and in spite of ample evidence for neural signal synchrony, especially in the form of gamma rhythms (a frequency range from about 35 to 90 hertz), the prediction that signals within sensory segments should be globally correlated has not been confirmed experimentally. This alone raises the question whether there are other mechanisms than signal synchrony by which the brain can express binding, and theory is called upon to work out proposals. (One such proposal for solving the segmentation problem without using temporal binding is described in [2].) And there is more work to do for theory. The above binding examples—attribute-object binding and sensory segmentation—are misleading in their simplicity, reducing the binding issue to the decomposition of the neural state into a few blocks, a view often defended by reference to our inability to keep simultaneously in mind more than a few chunks of a novel scene (the seven-plus-or-minus-two rule of [3]). On the other hand, we are evidently able to cope with very complex arrays of binding when representing a complex sentence, which necessitates to keep track simultaneously of multiple bindings between semantic, lexical, syntactic and phonetic elements, or when representing a visual scene of familiar structure, which necessitates the simultaneous handling of numerous relations between abstract and concrete patterns and their spatial relationships. Testimony to this complexity are the parsing trees of linguistics or the data structures of computer-based scene analysis (which themselves are all gross simplifications of the reality in our brains). Such complex relational patterns cannot be expressed by signal synchrony within realistic reaction times, given the poor temporal resolution of neural signals (1 to 3 msec, set by response times of neural membranes).

To do justice to the reality of our cognitive apparatus, we need a picture that lets us understand how the neural machinery in our head (or, for that matter, in a mouse's or salamander's head) is able to represent very intricate relational structures, and do so within typical reaction times of small fractions of a second. The called-for mechanisms must not only have high capacity and expressive power, but must in addition be able to store and retrieve relational structures once they have been formed. Finally, a clear picture must be developed for how the brain forms its preferred relational structures and how these preferred structures are to be characterized, for surely they can't be arbitrary.

A foreword is not the place to come forward with the proposal of a new system, but let me just remark that it is my conviction that rapid switching of synapses is part of the mechanism [4], and my laboratory has come to the conclusion that the machinery for storing and retrieving relational structures has the form of connections of a second order, of associative connections between switching synapses [5,6]. It is highly relevant to this book, however, to point out the fundamental significance of the time domain for these structures and processes, whatever they may be in detail.

To say it briefly, temporal signal structure is essential for expressing novel bindings, for laying down relational structures of growing complexity in memory, for reviving relational structures from memory (at a decisively reduced cost in terms of information rate) and for expressing bindings that resist memory storage. The mechanism for generating neural connectivity patterns, and, I claim, also of relational structures in memory, is network self-organization: the network creates structured activity patterns and synapses change in response to signal correlations, thus altering network and activity patterns. This reactive loop between network and activity tends to stabilize certain connectivity patterns, which are characterized by a close correspondence between signal correlations and connections. Network self-organization could perhaps be seen as a sequence of steps, each of which consists in the establishment of a temporal binding pattern followed by plastic change of connections, strengthening those between neurons bound to each other (that is, having correlated signals) while weakening those between neurons that are active but not bound to each other. Even if these individual binding patterns consist merely of one or a few blocks of bound neurons, the result of a sequence of such events can be a very intricate network of relations.

So far, network self-organization has been mostly applied to the generation of static networks, as illustrated by models of the ontogenesis of the visual system with its retinotopic connection patterns and columnar arrangements of sensory features (orientation, motion, stereo, color; for an example see [7]). If, however, synapses are allowed to switch on a fast time scale, a given set of neurons can support a number of alternate connectivity patterns, to be activated at different times. An important application of this could be neighborhood-preserving fiber projections corresponding to different transformation parameters to solve the problem of, for example, position-invariant pattern recognition [6]. For a model for how such alternate relational networks and their control structures could be generated by network self-organization, see [8].

Whereas the capacity of short-term memory is severely limited, as by Miller's seven-plus-or-minus-two rule, the capacity of long-term memory is generally held as virtually unlimited. The price to be paid is the laborious process of transferring short-term memory into long-term memory. Maybe this process is laborious because it necessitates the establishment of a new permanent relational network with the help of quite a number of consecutive activity binding patterns, as mentioned above.

Let me come back to our comparison between computer and brain. McCulloch and Pitts identified neurons with what in modern parlance are the logic gates—or bistable elements, or bits—of a digital machine. The bits of the computer can actually play the role of elements of pattern representations, analogous to the interpretation of neurons as elementary symbols. Many of them do, however, control switches (hence the name gate). Maybe it is time to reinterpret McCulloch and Pitts networks correspondingly, taking some of the "neurons" as elementary symbols, as is customary, but taking others as switches that can be opened and closed, an idea expressed already in [9].

The computer makes extensive use of temporal binding. All the bit settings in a given state are related to each other in the sense of forming one coherent functional state as specified in a program command. All signals necessary to constitute a state must have arrived at their target before the computer clock triggers the next state. The computer can afford this tight regime as its signals and pathways by now have a bandwidth of more than a gigahertz. In the brain, where the signal bandwidth is less than one kilohertz, a state comes into existence as the result of signals arriving without precise synchronization, so that the transition from one state to the next is a smooth and gradual affair.

The greatest step to be taken to transition from the computer to the brain is to find an explanation for the origin of states. As has been said above, whereas in the computer the switch settings essential for state organization are programmer-imposed, brain states must be self-organized. The gradual affair of brain state establishment may not just be a weakness but may be essential to this self-organization. If the brain has mechanisms to assess a state's level of self-consistency or completeness, it can iterate as long as it takes to establish a valid state. This complexity is the price the brain has to pay to be capable of programming itself as it goes along. If the state leaves behind a permanent trace that makes it easier to establish it, or parts of it, later again, and this self-programming may, after extensive exercise, install the equivalent of complex algorithms.

Unfortunately, our neural models are still very weak relative to this goal of brain state organization. This may be responsible for one great shortcoming of current neural network models and of related approaches—their inability to scale up in terms of numbers of elements or of functional sophistication to anything like the brains of even small animals. The difficulty is that larger systems cannot be made to converge to definite structures under the influence of training input. The solution to this problem must lie in decisive reduction of the systems' number of internal degrees of freedom, to be achieved by network self-organization (the one gigabyte of human genetic information not being enough to code for the petabyte needed to note down the wiring diagram of the human cortex). As an essential ingredient of any theory of network self-organization will be a clear understanding of the way in which temporal signal structure is shaped by a given network, the contents of this book seems to be highly relevant to neural network models of the coming decade.

References

1. McCulloch WS, Pitts W (1943) A logical calculus of ideas immanent in nervous activity. Bull Math Biophys 5:115–133
2. Wersing H, Steil JJ, Ritter HJ (2001) A competitive layer model for feature binding and sensory segmentation. Neural Comput 13(2):357–387. http://ni.www.techfak.uni-bielefeld.de/files/WersingSteilRitter2001-ACL.pdf
3. Miller GA (1956) The magical number seven, plus or minus two: some limits on our capacity for processing information. Psychol Rev 63:81–97
4. von der Malsburg C (1981) The correlation theory of brain function. Internal Report, 81-2, Max-Planck-Institut für Biophysikalische Chemie, Göttingen, Reprinted in Domany E, van Hemmen JL, Schulten K (eds) Models of neural networks II, Chap 2. Springer, Berlin, pp 95–119
5. Lücke J (2005) Information processing and learning in networks of cortical columns. Shaker Verlag, Dissertation
6. Wolfrum P, Wolff C, Lücke J, von der Malsburg C (2008) A recurrent dynamic model for correspondence-based face recognition. J Vis 8(7):1–18. http://journalofvision.org/8/7/34/, http://journalofvision.org/8/7/34/Wolfrum-2008-jov-8-7-34.pdf
7. Grabska-Barwinska A, von der Malsburg C (2009) Establishment of a Scaffold for orientation maps in primary visual cortex of higher mammals. J Neurosci 28:249–257. http://www.jneurosci.org/cgi/content/full/28/1/249
8. Bergmann U, von der Malsburg C (2010) A bilinear model for consistent topographic representations. In: Proceedings of ICANN, Part III, LNCS, vol 6354
9. Sejnowski TJ (1981) Skeleton filters in the brain. In: Hinton GE, Anderson JA (eds) Parallel models of associative memory. Lawrence Erlbaum, Hillsdale, pp 189–212

Frankfurt, Germany   Christoph von der Malsburg

Acknowledgements

We are delighted to bring out a book dedicated to understanding the role of timing information in brain function. This has proven to be a daunting challenge. However, with the aid of advanced neuroscientific measurement techniques, more sophisticated mathematical modeling techniques, increased computational power and fast hardware implementations, we are making rapid progress.

We are very grateful to the contributing authors of the various chapters in the book for their valuable insights. We are particularly delighted to receive a Foreword written by Dr. Christoph von der Malsburg, a pioneer in this field.

We appreciate the efficient publication services provided by Ann Avouris and her staff at Springer. We are also grateful to the management at IBM Research, specifically Dr. Charles Peck and Dr. Ajay Royyuru in the Computational Biology Center for their support of this publication project.

A.R. Rao
G.A. Cecchi

Contents

1 **Introduction** 1
   Guillermo Cecchi and A. Ravishankar Rao

2 **Adaptation and Contraction Theory for the Synchronization of Complex Neural Networks** 9
   Pietro DeLellis, Mario di Bernardo, and Giovanni Russo

3 **Temporal Coding Is Not Only About Cooperation—It Is Also About Competition** 33
   Thomas Burwick

4 **Using Non-oscillatory Dynamics to Disambiguate Pattern Mixtures** . 57
   Tsvi Achler

5 **Functional Constraints on Network Topology via Generalized Sparse Representations** 75
   A. Ravishankar Rao and Guillermo A. Cecchi

6 **Evolution of Time in Neural Networks: From the Present to the Past, and Forward to the Future** 99
   Ji Ryang Chung, Jaerock Kwon, Timothy A. Mann, and Yoonsuck Choe

7 **Synchronization of Coupled Pulse-Type Hardware Neuron Models for CPG Model** 117
   Ken Saito, Akihiro Matsuda, Katsutoshi Saeki, Fumio Uchikoba, and Yoshifumi Sekine

8 **A Universal Abstract-Time Platform for Real-Time Neural Networks** 135
   Alexander D. Rast, M. Mukaram Khan, Xin Jin, Luis A. Plana, and Steve B. Furber

9 **Solving Complex Control Tasks via Simple Rule(s): Using Chaotic Dynamics in a Recurrent Neural Network Model** 159
   Yongtao Li and Shigetoshi Nara

10 Time Scale Analysis of Neuronal Ensemble Data Used to Feed Neural Network Models 179
N.A.P. Vasconcelos, W. Blanco, J. Faber, H.M. Gomes, T.M. Barros, and S. Ribeiro

11 Simultaneous EEG-fMRI: Integrating Spatial and Temporal Resolution 199
Marcio Junior Sturzbecher and Draulio Barros de Araujo

Index 219

Contributors

Tsvi Achler Siebel Center, University of Illinois Urbana-Champaign, 201 N. Goodwin Ave, Urbana, IL 61801, USA, achler@illinois.edu

Draulio Barros de Araujo Department of Physics, FFCLRP, University of Sao Paulo, Ribeirao Preto, Brazil, draulio@neuro.ufrn.br; Brain Institute, Federal University of Rio Grande do Norte, Natal, Brazil; Onofre Lopes University Hospital, Federal University of Rio Grande do Norte, Natal, Brazil

T.M. Barros Edmond and Lily Safra International Institute of Neuroscience of Natal (ELS-IINN), Rua Professor Francisco Luciano de Oliveira 2460, Bairro Candelária, Natal, RN, Brazil

W. Blanco Brain Institute, Federal University of Rio Grande do Norte (UFRN), Natal, RN, 59078-450, Brazil

Thomas Burwick Frankfurt Institute for Advanced Studies (FIAS), Goethe-Universität, Ruth-Moufang-Str. 1, 60438 Frankfurt am Main, Germany, burwick@fias.uni-frankfurt.de

Guillermo Cecchi IBM Research, Yorktown Heights, NY 10598, USA, gcecchi@us.ibm.com

Yoonsuck Choe Department of Computer Science and Engineering, Texas A&M University, 3112 TAMU, College Station, TX 77843-3112, USA, choe@cs.tamu.edu

Ji Ryang Chung Department of Computer Science and Engineering, Texas A&M University, 3112 TAMU, College Station, TX 77843-3112, USA, jchung@cse.tamu.edu

Pietro DeLellis Department of Systems and Computer Engineering, University of Naples Federico II, Naples, Italy, pietro.delellis@unina.it

Mario di Bernardo Department of Systems and Computer Engineering, University of Naples Federico II, Naples, Italy, mario.dibernardo@unina.it; Department of Engineering Mathematics, University of Bristol, Bristol, UK, m.dibernardo@bristol.ac.uk

xv

J. Faber Fondation Nanosciences & CEA/LETI/CLINATEC, Grenoble, 38000, France

Steve B. Furber School of Computer Science, University of Manchester, Manchester, UK M13 9PL, steve.furber@manchester.ac.uk

H.M. Gomes Department of Systems and Computation, Federal University of Campina Grande (UFCG), Campina Grande, PB, 58249-900, Brazil

Xin Jin School of Computer Science, University of Manchester, Manchester, UK M13 9PL, jinxa@cs.man.ac.uk

M. Mukaram Khan School of Computer Science, University of Manchester, Manchester, UK M13 9PL, khanm@cs.man.ac.uk

Jaerock Kwon Department of Electrical and Computer Engineering, Kettering University, 1700 W. University Avenue, Flint, MI 48504, USA, jkwon@kettering.edu

Yongtao Li Department of Electrical & Electronic Engineering, Graduate School of Natural Science and Technology, Okayama University, Okayama, Japan, yongtaoli@es.hokudai.ac.jp

Timothy A. Mann Department of Computer Science and Engineering, Texas A&M University, 3112 TAMU, College Station, TX 77843-3112, USA, mann@cse.tamu.edu

Akihiro Matsuda College of Science and Technology, Nihon University, 7-24-1 Narashinodai, Funabashi-shi, Chiba, 274-8501 Japan

Shigetoshi Nara Department of Electrical & Electronic Engineering, Graduate School of Natural Science and Technology, Okayama University, Okayama, Japan, nara@chaos.elec.okayama-u.ac.jp

Luis A. Plana School of Computer Science, University of Manchester, Manchester, UK M13 9PL, plana@cs.man.ac.uk

A. Ravishankar Rao IBM Research, Yorktown Heights, NY 10598, USA, ravirao@us.ibm.com

Alexander D. Rast School of Computer Science, University of Manchester, Manchester, UK M13 9PL, rasta@cs.man.ac.uk

S. Ribeiro Brain Institute, Federal University of Rio Grande do Norte (UFRN), Natal, RN, 59078-450, Brazil, sidartaribeiro@neuro.ufrn.br; Neuroscience Graduate Program, Federal University of Rio Grande do Norte (UFRN), Natal, RN, 59078-450, Brazil; Psychobiology Graduate Program, Federal University of Rio Grande do Norte (UFRN), Natal, RN, 59078-450, Brazil

Giovanni Russo Department of Systems and Computer Engineering, University of Naples Federico II, Naples, Italy, giovanni.russo2@unina.it

Katsutoshi Saeki College of Science and Technology, Nihon University, 7-24-1 Narashinodai, Funabashi-shi, Chiba, 274-8501 Japan

Ken Saito College of Science and Technology, Nihon University, 7-24-1 Narashinodai, Funabashi-shi, Chiba, 274-8501 Japan, kensaito@eme.cst.nihon-u.ac.jp

Yoshifumi Sekine College of Science and Technology, Nihon University, 7-24-1 Narashinodai, Funabashi-shi, Chiba, 274-8501 Japan

Marcio Junior Sturzbecher Department of Physics, FFCLRP, University of Sao Paulo, Ribeirao Preto, Brazil

Fumio Uchikoba College of Science and Technology, Nihon University, 7-24-1 Narashinodai, Funabashi-shi, Chiba, 274-8501 Japan

N.A.P. Vasconcelos Brain Institute, Federal University of Rio Grande do Norte (UFRN), Natal, RN, 59078-450, Brazil; Department of Systems and Computation, Federal University of Campina Grande (UFCG), Campina Grande, PB, 58249-900, Brazil; Edmond and Lily Safra International Institute of Neuroscience of Natal (ELS-IINN), Rua Professor Francisco Luciano de Oliveira 2460, Bairro Candelária, Natal, RN, Brazil; Faculdade Natalense para o Desenvolvimento do Rio Grande do Norte (FARN), Natal, RN 59014-540, Brazil; Faculdade de Natal, Natal, RN, 59064-740, Brazil

Chapter 1
Introduction

Guillermo Cecchi and A. Ravishankar Rao

Abstract The field of neural modeling uses neuroscientific data and measurements to build computational abstractions that represent the functioning of a neural system. The timing of various neural signals conveys important information about the sensory world, and also about the relationships between activities occurring in different parts of a brain. Both theoretical and experimental advances are required to effectively understand and model such complex interactions within a neural system. This book aims to develop a unified understanding of temporal interactions in neural systems, including their representation, role and function. We present three different research perspectives arising from theoretical, engineering and experimental approaches.

A significant amount of effort in neural modeling is directed towards understanding the representation of external objects in the brain, prominently in primary and associative cortical areas, and along the pathways that process sensory information from the periphery. There is also a rapidly growing interest in modeling the intrinsically generated activity in the brain represented by the default mode state, the emergent behavior that gives rise to critical phenomena such as neural avalanches, and the self-generated activity required to drive behavior. Time plays a critical role in these intended modeling domains, from the mundane yet exquisite discriminations the mammalian auditory system achieves in echolocation and voice recognition, to the precise timing involved in high-end activities such as competitive sports or professional music performance.

The effective incorporation of time in neural network models, however, has been a challenging task. Inspired by early experimental observations of oscillatory activity in electro-encephalogram recordings, and more recently in magneto-encephalogram observations [9], many theoretical efforts have been focused on the emergence and functionality of oscillations and synchronized activity of neural pop-

G. Cecchi (✉) · A.R. Rao
IBM Research, Yorktown Heights, NY 10598, USA
e-mail: gcecchi@us.ibm.com

A.R. Rao
e-mail: ravirao@us.ibm.com

ulations. In fact, the phenomenon of synchronization of ensembles of oscillators has been long recognized as an essential feature of biological systems. The pioneering work of Winfree and Kuramoto [10, 15] laid the foundation for a theoretical analysis of oscillator networks with relatively simple configurations. In neuroscience, similarly, Wilson and Cowan provided a framework to analyze the conditions under which ensembles of locally interacting neurons can give rise to oscillations. These early ideas have permitted researchers to find a conceptual link between neural modeling and the variety of complex perceptual and cognitive brain states that require local or long-distance coordination of neural ensembles, such as perceptual binding of real and illusory contours [5], face recognition [12], sensorimotor tasks [8], and attention [4], for which experimental evidence of involvement of oscillations has been documented.

Despite this wealth of observations, theoretical frameworks to conceptualize the full functional implications of oscillatory networks remain scattered and disconnected. Moreover, most modeling efforts are descriptive or heuristic, and tailored to specific, constrained situations. The purpose of this book is to provide a rallying point to develop a unified view of how the time domain can be effectively employed in neural network models. We will concentrate on three broad lines of research that run in parallel, but have enough sleepers to bind them together.

A first direction to consider is the utilization of ensembles of oscillators with the purpose of achieving specific, well-designed computational tasks. This is exemplified by the use of synchronization between oscillators as a means to solve the binding problem, that is, how to carve out all the units that, across different levels of abstraction, contribute to the emergence of a perceptual or cognitive object, without including other unrelated active units and without requiring a combinatorial, exploding number of connections. This issue, with implications for several related problems in neuroscience and signal processing such as perceptual integration and image segmentation, has been the focus of intense research in recent years, including approaches from biological, engineering and mathematical perspectives that range from physiological plausibility to adaptive network topologies for synchronization.

Motivated mostly by the need to implement practical applications for signal processing, a second line of interest is the development of dedicated hardware solutions for fast, real-time simulations of large-scale networks of oscillators. The implications of this type of research are enormous, and certainly go beyond the application to signal processing, as a successful hardware simulator could in principle form the basis on which to expand and multiply recent advances in the field of neural implants, neuro-prosthetics and brain-machine interfaces.

Finally, the advancement of electro-physiology techniques, including the development of new ones, together with practically inexhaustible computational power, has allowed for increased availability of experimental data to match the different levels of abstraction, be they multi-electrode recordings, electro- and magneto-encephalo- and cortico-grams (EEG, MEG, ECoG), functional magnetic resonance (fMRI), and ultra-fast optical imaging [1, 2, 6, 7, 11]. As a consequence, a third and in fact enormously active direction of research is the extraction of temporal patterns from the various sources of brain data, with objectives that range from the theoretically motivated to purely clinical studies.

1 Introduction

The chapters have been organized accordingly, reflecting these three research approaches: theoretical, engineering and experimental. Following is a brief description of each contribution in relation to the objectives of the book.

1.1 Theoretical Background

While the synchronization of relaxation oscillators in simple network topologies is well understood, extensions to less constrained conditions are still an open area of research. In Chap. 2, De Lellis, di Bernardo and Russo investigate the conditions that allow synchronization in dynamical systems consisting of networks of units with arbitrary dynamics, topologies and coupling functions. They explore two parallel approaches. First, they focus on network plasticity: under very general assumptions on the vector field that defines the dynamics of the neural units (the QUAD condition, a generalization of Lipschitz's condition for uniqueness of solution of ordinary differential equations), it is possible to derive conditions for the dynamics of weight adaptation that ensure synchronization. Interestingly, the resulting update rule looks like a generalization of Hebbian learning, providing a plausible path for physiological interpretation. Alternatively, the authors ask whether it is possible to specify the node dynamics such that under simple coupling functions, and without the need of topology or weight adaptation, synchronization will be guaranteed. They answer this question through the use of contraction theory, which defines exponentially decaying bounds for the divergence of nearby trajectories of a dynamical system, if the norm-induced measure of the Jacobian is appropriately bound over the entire phase space. With this result, they show that the simple addition of self-inhibitory feedback to the units results in asymptotically synchronizing networks, for a large and simple class of coupling functions.

One of the most widely studied computational applications of neural synchrony is as a solution to the superposition catastrophe problem [13]. In Chap. 3, Burwick proposes an intriguing alternative to typical phase-locking synchrony implementations. One manifestation of the superposition problem is that when a network learns to recognize patterns with substantial overlap at the input level (e.g., many pixels in common), a fine-tuned layer of inhibitory units is required to avoid the spread of activation, and synchronization, to all of the units when any given pattern is presented.

Instead of relying on inhibition and units oscillating at constant frequency (at least in steady-state), the author shows that an acceleration term added to the dynamics of the units has the effect of creating a coherent synchronization only for the units that encode the winning pattern; the other units, while active, fail to synchronize, and instead perform a precessional, non-stationary motion in the space of phases. The acceleration term depends only on the lateral connections between the units, and is responsible for the winning ensemble "leaving behind" the units that do not sufficiently recognize the pattern.

A drawback of oscillatory models is that it is difficult to observe patterns of activation in individual neurons consistent with the theory; oscillations seem to truly be a collective phenomenon. As a consequence, for synchronization to take place, oscillatory networks require time in order to transmit phase information back and forth across the ensemble, limiting their computational capacity as well as their modeling scope. In Chap. 4, Achler proposes a mechanism to address the binding of such distributed representations, which is achieved by feedback inhibition, rather than oscillations. The recurring inhibitory feedback suppresses the activity of "distractors" that do not belong to the object-encoding ensemble. While it still requires time for the recurrence to play out, this mechanism may potentially result in faster convergence times, and furthermore provide an alternative physiologically plausible model for binding.

While optimization approaches are common in signal processing, computer vision and, under various guises, in theories of efficient coding for early sensory processing, little progress has been made towards generalizing optimization principles to the realm of oscillatory networks. In Chap. 5, Rao and Cecchi develop an approach towards understanding sensory coding in an oscillatory network based on maximizing the sparseness of the spatio-temporal representation of inputs. This leads to a network dynamics by which higher-level representations of object features are synchronized with the lower-level object primitives. Furthermore, a superposition of input objects produces higher-level representations that preserve the distinction between the objects via the phases of the oscillations. This behavior leads to a quantitative characterization of the network behavior in terms of its efficacy in classifying and disambiguating superposed objects. These quantitative measures of network behavior are a function of the network topology, and depend on the fan-out of feed-forward, feedback and lateral connections. Rao and Cecchi show that these quantitative measures of network behavior are maximized when the network topology is qualitatively similar to the topology of brain networks.

1.2 Engineering Development and Applications

Recollection and prediction constitute two essential attributes of behavior in higher-order organisms such as vertebrate animals. In Chap. 6, Chung et al. demonstrate that these attributes can arise through neural network controllers embedded in a dynamic environment. Both recollection and prediction require a temporal comparison in that the state of the organism or environment in the past or future is compared against the current state. In this sense, temporal dynamics are essential in order to represent interactions between the organism and environment. The authors use two tasks, that of ball-catching and pole-balancing to illustrate their framework for representing recollection and prediction. They use classical feed-forward and recurrent networks, and a hierarchy of retro- and predictive configurations based on the access to external markers and to internal dynamical regularities. Their work provides a possible framework to understand the representation of temporal order in the brain.

Overcoming the computational bottleneck is a major challenge facing researchers interested in exploring temporal phenomena in neural networks. This problem arises because simulations of these networks need to be carried out iteratively over small time increments in order to preserve numerical stability and the accuracy of the

1 Introduction

simulations. This implies that several iterations may be necessary to produce oscillations and phenomena such as synchronization. Hence, hardware implementations of the underlying models are particularly attractive, as real-time behavior can be achieved.

In Chap. 7, Saito et al. present hardware-based neural models that reproduce the biologically observed characteristics of neural responses. They offer several improvements over existing hardware implementations, such as eliminating the need for inductors, thereby enabling a realization using CMOS IC chips. Their system has been implemented and tested to demonstrate the generation of oscillatory patterns that govern locomotion in micro-robots.

In Chap. 8, Rast et al. describe an alternative method to address direct hardware implementations of neural network systems. They have successfully designed a neural chip multiprocessor called SpiNNaker. This platform allows users to specify network models using a high-level hardware description language. The authors have implemented spiking neuron models with support for asynchronous interactions between the neurons. Their goal is to be able to simulate networks with a billion neurons, which approaches the size of a mammalian brain. This research constitutes a promising exploration of the challenges in designing a highly scalable architecture for performing simulations of large neural networks.

The ultimate goal of much of the research in neural systems modeling is to be able to produce behavior that mimics the function of biological organisms, particularly in the way real brains operate. Several researchers work on component models that explain the functioning of specific pieces of the entire puzzle, such as the encoding and processing of visual sensory information, or motor control. However, there are relatively few efforts to integrate such component models together. In Chap. 9, Li and Nara present a roving robot that combines sensory processing with motor behavior in order to solve a maze navigation task. Such a task is ill-posed, and requires appropriate control rules to be able to navigate around obstacles in order to reach a desired destination. Li and Nara use the mechanism of chaotic dynamics to implement an effective navigation scheme. The desired destination is signaled by means of auditory signals. When confronted with an obstacle, their system is able to generate alternate paths to reach the destination.

1.3 Biological Experimentation

As already stated, one of the main goals of the inclusion of time in neural network is to create a theoretical framework to understand the behavior of the nervous system. In Chap. 10, Vasconcelos et al. ask the experimental counterpart to the theories of temporal processing: is there a codification of external objects that specifically relies on timing? They utilize a multi-electrode technique pioneered by Miguel Nicolelis and John Chapin that allows the simultaneous recording of the electrophysiological activity of up to hundreds of neurons, distributed across several cortical and sub-cortical areas of the rat's brain. With this setup, they show that different objects with which the rats interact display a unique activity pattern, when the activity is represented by the average firing rate within time bins with sizes ranging from 40 to 1,000 msec. Interestingly, they find that the optimal temporal resolution and the relative contribution of the different brain areas, in terms of object coding, are object-dependent, suggesting that the intrinsic features of the objects, as well as the behavioral interaction patterns the animals engage in with the, determine the coding features.

Moreover, they also find that exposing the animals to the objects and allowing the interact with them increases the correlation between the firing patterns with those generated in the subsequent periods of slow-wave and REM sleep, a finding that parallels those obtained by Bruce McNaughton in the hippocampus and Dan Margoliash in the songbird system [3, 14].

In Chap. 11, Sturzbecher and de Araujo expound on the possibilities and limitations of combining simultaneous recordings of functional magnetic resonance imaging (fMRI) and electro-encephalograms (EEG), from the perspective of increasing the spatial and temporal resolution of functional data. Focusing on the practical application of identifying the brain region from which seizures originate in epileptic patients (the epileptogenic zone), they explain that the problem of colocating EEG and fMRI sources can be addressed by solving the surrogate problem of mapping the traces created by the very frequent interictal epileptic discharges (IED), typical of the disease even in the absence of seizures. The inference of fMRI/EEG source location is normally based on the general linear model (GLM), which assumes a fixed temporal profile for the electro-physiology to hemodynamic response function, and measures the amplitude of the fMRI response to the IED's using linear correlation. The authors show that by using Kullback–Leibler divergence to measure these responses, which does not require any assumption about the transfer response, it is possible to significantly increase the accuracy of source localization. Their finding illustrates a concrete, practical application of an explicit inclusion of time in the analysis of real neural data.

References

1. Berdyyeva TK, Reynolds JH (2009) The dawning of primate optogenetics. Neuron 62(2):159–160
2. Cardin JA et al (2009) Driving fast-spiking cells induces gamma rhythm and controls sensory responses. Nature 459(7247):663–667
3. Dave AS, Margoliash D (2000) Song replay during sleep and computational rules for sensorimotor vocal learning. Science 290:812–816
4. Fries P et al (2001) Modulation of oscillatory neuronal synchronization by selective visual attention. Science 291(5508):1560–1563
5. Gray CM et al (1989) Oscillatory responses in cat visual cortex exhibit inter-columnar synchronization which reflects global stimulus properties. Nature 338:334–337
6. Haider B, McCormick DA (2009) Rapid neocortical dynamics: cellular and network mechanisms. Neuron 62(2):171–189
7. Han X et al (2009) Millisecond-timescale optical control of neural dynamics in the nonhuman primate brain. Neuron 62(2):191–198
8. Jefferys JGR, Traub RD, Whittington MA (1996) Neuronal networks for induced '40 Hz' rhythms. Trends Neurosci 19:202–208

1 Introduction

9. Joliot M, Ribary U, Llinás R (1994) Human oscillatory brain activity near 40 Hz coexists with cognitive temporal binding. Proc Natl Acad Sci USA 91(24):11748–11751
10. Kuramoto Y (1984) Chemical oscillations, waves, and turbulence. Springer, Berlin
11. Liu H et al (2009) Timing, timing, timing: fast decoding of object information from intracranial field potentials in human visual cortex. Neuron 62(2):281–290
12. Rodriguez E et al (1999) Perception's shadow: long-distance synchronization of human brain activity. Nature 397:430–433
13. von der Malsburg C (1999) The what and why of binding: the modeler's perspective. Neuron 24:95–104
14. Wilson MA, McNaughton BL (1994) Reactivation of hippocampal ensemble memories during sleep. Science 265:676–679
15. Winfree AT (2001) The geometry of biological time, 2nd edn. Springer, Berlin

Chapter 2
Adaptation and Contraction Theory for the Synchronization of Complex Neural Networks

Pietro DeLellis, Mario di Bernardo, and Giovanni Russo

Abstract In this chapter, we will present two different approaches to solve the problem of synchronizing networks of interacting dynamical systems. The former will be based on making the coupling between agents in the network adaptive and evolving so that synchronization can emerge asymptotically. The latter will be using recent results from contraction theory to give conditions on the node dynamics and the network topology that result into the desired synchronized motion. The theoretical results will be illustrated by means of some representative examples, including networks of neural oscillators.

2.1 Introduction

Synchronization and coordination of motion are key features of many biological and living systems. For example, circadian rhythms regulate the functions of cells in our bodies which are entrained to the day/night cycle via the sophisticated action of various gene regulatory networks, see, for example, [14]. More notably, synchronization has been proposed as a powerful mechanism to explain some of the patterns observed in the brain [1, 3, 19, 27]. For instance, as noted in [18], partial synchrony in cortical networks can be used to explain several brain oscillatory patterns such as the alpha and gamma EEG rhythms. Also, synchronization of brain waves has

---

P. DeLellis · M. di Bernardo · G. Russo
Department of Systems and Computer Engineering, University of Naples Federico II, Naples, Italy

P. DeLellis
e-mail: pietro.delellis@unina.it

M. di Bernardo
e-mail: mario.dibernardo@unina.it

G. Russo
e-mail: giovanni.russo2@unina.it

M. di Bernardo (✉)
Department of Engineering Mathematics, University of Bristol, Bristol, UK
e-mail: m.dibernardo@bristol.ac.uk

been proposed to explain pathologies such as epilepsy while coordinated neuronal evolution is required to generate patterns essential for locomotion, breathing, etc.

In all of these instances, sets of agents (genes, neurons etc.) communicate with each other over a complex web of interconnections exchanging information on their mutual state. Through this distributed communication process, agents successfully negotiate how to change their own behavior so that some synchronous evolution emerges. Recent studies in synchronization have clearly shown that such a process is, in general, completely decentralized with no central supervisor determining how individual agents should behave. This fascinating feature of synchronization makes it possible for it to be observed also in those cases where uncertainties and external disturbances can affect the agent behavior. Indeed, networks of agents have been shown to achieve synchronization in a plethora of different ways under all sort of different conditions, for example, [14, 16, 17, 19, 22, 23, 37].

Usually in the literature to model and investigate synchronization phenomena, a network of interacting dynamical agents is often considered. Specifically, let $\mathscr{G} = \{\mathscr{N}, \mathscr{E}\}$ be a graph defined by the set of nodes $\mathscr{N}$ and the set of edges $\mathscr{E}$, having cardinality N and M, respectively. The classical model that has been used to describe a complex network of N nodes is

$$\dot{x}_i = f(x_i, t) - \sigma \sum_{j=1}^{N} \ell_{ij} h(x_j), \quad i = 1, \ldots, N. \tag{2.1}$$

Here, $x_i \in \mathbb{R}^n$ is the state of node i, $f : \mathbb{R}^n \times \mathbb{R}^+ \to \mathbb{R}^n$ is the vector field describing the dynamics of an isolated node, $\sigma \in \mathbb{R}$ is the coupling strength, $h : \mathbb{R}^n \to \mathbb{R}^n$ is the output function through which the network nodes are coupled, and ℓ_{ij} is the ij-th element of the Laplacian matrix $\mathscr{L}$ associated to the graph $\mathscr{G}$ describing the network topology.

Several simplifying assumptions are often made in the literature in order to study the problem. For example, it is typically assumed that all nodes share the same dynamics, that the network topology is time-invariant and that a unique coupling gain, σ, determines the strength of the coupling between neighboring nodes. Clearly, these assumptions are unrealistic when compared to the natural world. Indeed in Nature, the coupling strength between agents as well as the very structure of their interconnections vary in time so that the network can adapt and evolve in order to maintain a synchronous evolution. Moreover, often in biological networks evolution leads to systems and coupling laws whose structure facilitate the emergence of synchronization.

In this chapter, we discuss two approaches to analyze and induce synchronization in networks of dynamical agents with the aim of mimicking these features of the natural world. Specifically, we will present a strategy to adapt and evolve a given network of dynamical agents so as to make all agents synchronize onto a common evolution. Also, we will show that by using contraction theory, a tool from the theory of dynamical systems, it is possible to analyze and design the agent behavior and coupling so that the network spontaneously synchronizes. We will validate both strategies on a network of FitzHugh–Nagumo neurons showing that indeed they are effective methodologies to guarantee their synchronization. This is, of course, only a first step in showing that adaptation and evolution can play an important role in determining the effectiveness of those synchronization processes commonly detected in natural systems and the brain.

The rest of the chapter is outlined as follows. We first present in Sect. 2.2 the basic mathematical tools used for deriving our results. In Sect. 2.3, we present two synchronization strategies based on the use of adaptive coupling strengths. In Sect. 2.4, we then turn our attention to the problem of tuning some properties of neurons so that synchronization spontaneously emerges, while in Sect. 2.4.1 we present the main tool used to this aim. Finally, in Sect. 2.5 and Sect. 2.6 we illustrate our theoretical results and validate them for networks of FitzHugh–Nagumo neurons. Conclusions are drawn in Sect. 2.8.

2.2 Preliminaries

We consider systems of ordinary differential equations, generally time-dependent:

$$\dot{x} = f(t, x) \qquad (2.2)$$

defined for $t \in [0, \infty)$ and $x \in C$, where C is a subset of $\mathbb{R}^n$. It will be assumed that $f(t, x)$ is differentiable on x, and that $f(t, x)$, as well as the Jacobian of f with respect to x, denoted as $J(t, x) = \frac{\partial f}{\partial x}(t, x)$, are both continuous in (t, x). In applications of the theory, it is often the case that C will be a closed set, for example given by non-negativity constraints on variables as well as linear equalities representing mass-conservation laws. For a nonopen set C, differentiability in x means that the vector field $f(t, \bullet)$ can be extended as a differentiable function to some open set which includes C, and the continuity hypotheses with respect to (t, x) hold on this open set.

The solution of Eq. 2.2 is in principle defined only on some interval $s \le t < s + \varepsilon$, but we will assume that it is defined for all $t \ge s$. Conditions which guarantee such a property are often satisfied in biological applications, for example, whenever the set C is closed and bounded, or whenever the vector field f is bounded. (See Appendix C in [36] for more discussion, as well as [2] for a characterization of the forward completeness property.)

One of the main assumptions used in literature to prove asymptotic synchronization of network Eq. 2.1 is the so-called QUAD assumption, which can be defined as follows:

Definition 2.1 A function $f : \mathbb{R}^n \times \mathbb{R} \mapsto \mathbb{R}^n$ is QUAD(Δ, ω) if and only if, for any $x, y \in \mathbb{R}^n, t \in \mathbb{R}^+$:

$$(x - y)^T \left[f(x, t) - f(y, t) \right] - (x - y)^T \Delta (x - y) \le -\omega (x - y)^T (x - y), \qquad (2.3)$$

where Δ is an $n \times n$ diagonal matrix and ω is a real positive scalar.

Table 2.1 Standard matrix measures used in this work

Vector norm, $|\cdot|$	Induced matrix measure, $\mu(A)$
$\|x\|_1 = \sum_{j=1}^m \|x_j\|$	$\mu_1(A) = \max_j(a_{jj} + \sum_{i \neq j}\|a_{ij}\|)$
$\|x\|_2 = (\sum_{j=1}^n \|x_j\|^2)^{\frac{1}{2}}$	$\mu_2(A) = \max_i(\lambda_i\{\frac{A+A^*}{2}\})$
$\|x\|_\infty = \max_{1 \leq j \leq m}\|x_j\|$	$\mu_\infty(A) = \max_i(a_{ii} + \sum_{j \neq i}\|a_{ij}\|)$

Another possibility is to make use of contraction theory, a powerful tool from the theory of dynamical systems aimed at assessing the stability between trajectories in the state space of a system or network of interest. The idea is to assess under what conditions all trajectories of a given network or system converge towards each other and hence onto some common asymptotic evolution. In order to properly define this concept, we recall (see, for instance, [24]) that, given a vector norm on Euclidean space ($|\bullet|$), with its induced matrix norm $\|A\|$, the associated *matrix measure* μ is defined as the directional derivative of the matrix norm, that is,

$$\mu(A) := \lim_{h \searrow 0} \frac{1}{h}\big(\|I + hA\| - 1\big).$$

For example, if $|\bullet|$ is the standard Euclidean 2-norm, then $\mu(A)$ is the maximum eigenvalue of the symmetric part of A. As we shall see, however, different norms will be useful for our applications. In particular, we will make use of the 1-norm and ∞-norm, which are reported in Table 2.1.

We say that a system such as the one in Eq. 2.2 is (*infinitesimally*) *contracting* [21] on a convex set $C \subseteq \mathbb{R}^n$ if there exists some norm in C, with associated matrix measure μ such that, for some constant $c \in \mathbb{R} - \{0\}$,

$$\mu\big(J(x,t)\big) \leq -c^2, \quad \forall x \in C, \ \forall t \geq 0. \tag{2.4}$$

The key theoretical result about contracting systems links infinitesimal and global contractivity, and is stated below. This result was presented in [21] and historically can be traced, under different technical assumptions, to for example, [15, 20, 26].

Theorem 2.1 *Suppose that C is a convex subset of $\mathbb{R}^n$ and that $f(t, x)$ is infinitesimally contracting with contraction rate c^2. Then, for every two solutions $x(t)$ and $z(t)$ rooted in $\xi \in C$ and $\zeta \in C$ respectively, of Eq. 2.2, it holds that:*

$$\big|x(t) - z(t)\big| \leq e^{-c^2 t}|\xi - \zeta|, \quad \forall t \geq 0. \tag{2.5}$$

2.3 Adaptive Synchronization of Complex Networks

In the classical description of a complex network, see Eq. 2.1, the intensity of the coupling σ between nodes is often assumed to be constant and equal for all pairs of interconnected nodes. Following [5–7], it is possible to consider a different coupling strategy where neighboring nodes negotiate the strength of their mutual coupling, effectively adapting its magnitude as a function of their mismatch. The network equation equipped with such a decentralized gain adaptation law can be written as

$$\dot{x}_i = f(x_i, t) + \sum_{j \in \mathscr{E}_i} \sigma_{ij}(t)\big(h(x_j) - h(x_i)\big), \qquad (2.6)$$

$$\dot{\sigma}_{ij} = g\big(h(x_j) - h(x_i)\big), \qquad (2.7)$$

where $\sigma_{ij}(t)$ is the coupling gain associated to the generic edge (i, j) between node i and node j and $g : \mathbb{R}^n \mapsto \mathbb{R}$ is some scalar function of the output error between neighboring nodes which is used to adapt and evolve the value of the gain itself.

Obviously, the effectiveness of the adaptation process is closely linked to the choice of the function g which is fundamental in determining the occurrence and performance of the synchronization process on the network. Here, we propose two possible alternative local strategies to guarantee the emergence of a synchronous evolution. The *vertex-based* strategy where the same adaptive gain is associated with all edges outgoing from a given node; and the *edge-based* strategy, where, instead, an adaptive gain is associated with every link in the network. We now briefly describe each of the two strategies.

- **Vertex-based strategy**. A first possible way of introducing gain adaptation is to associate an adaptive gain $\sigma_i(t)$ to every vertex i in the network. The gains associated to the outgoing links of each node are then updated comparing its output with the output of its neighbors, see [7]. Namely, the adaptation law is chosen as:

$$\dot{\sigma}_{ij}(t) = \dot{\sigma}_i(t) = \eta \left\| \sum_{j \in \mathscr{E}_i} \big(h(x_j) - h(x_i)\big) \right\|, \quad \mu > 0, \ (i, j) \in \mathscr{E}, \qquad (2.8)$$

where η is some scalar parameter to be chosen as part of the design process, determining how sensitive the gain variation is to changes in the error between the node outputs. An alternative vertex-based strategy was independently proposed by Zhou and Kurths in [40] and is characterized by upper-bounding the maximal rate of change of the coupling gain.
- **Edge-based strategy**. Another possible decentralized approach to tune the coupling gains is to adaptively update each non-null element of the adjacency matrix. In other words, an adaptive gain σ_{ij} is associated with every $(i, j) \in \mathscr{E}$ [7]. We propose the following adaptation law:

$$\dot{\sigma}_{ij}(t) = \alpha \|h(x_j) - h(x_i)\|, \quad \alpha > 0, \ (i, j) \in \mathscr{E}. \qquad (2.9)$$

Adopting this strategy, each couple of nodes negotiates the strength of their mutual coupling on the basis of the distance between their outputs.

The stability of these adaptive strategies was investigated in [6]. It was shown that, assuming $h(x) = x$, that is, all the nodes are linearly coupled through all their state variables, a sufficient condition for asymptotic synchronization is that the vector field describing the node dynamics is QUAD(Δ, ω) with $\Delta - \omega I \leq 0$. In [5], the stability analysis was extended to a more general class of adaptive laws. Namely, the adaptive function belongs to one of the two following classes:

- Class 1: $g(e_{ij}) = \alpha \|e_{ij}\|^p$, with $e_{ij} = x_j - x_i$, $0 < p \leq 2$.
- Class 2: $g(e_{ij}) = \kappa(\|e_{ij}\|)$, where κ is a bounded differentiable class k function[1].

Theorem 2.2 *Let us assume that f is a continuous and differentiable vector field which is also QUAD(Δ, ω) with $\Delta - \omega I \leq 0$, and that $h(x) = x$. If the function g is selected from classes 1 or 2, then the network in Eq. 2.6, Eq. 2.7 asymptotically synchronizes onto a common evolution with*

$$\lim_{t \to \infty} (x_j - x_i) = 0, \quad \forall (i, j) \in \mathcal{E}, \tag{2.10}$$

and

$$\lim_{t \to \infty} \sigma_{ij}(t) = c_{ij} < +\infty, \quad \forall (i, j) \in \mathcal{E}. \tag{2.11}$$

(See [5] for the proof which is based on the use of appropriate Lyapunov functions to show that all nodes converge asymptotically towards the same synchronous evolution.)

This theorem ensures the asymptotic stability and the boundedness of the adaptive gains for any connected network, under the assumption that the vector field is QUAD. Nonetheless, in neural networks the coupling is typically on only one state variable (see, for instance, [25]), and, moreover, the vector field describe the node dynamics is QUAD, but $\Delta - \omega I$ is not negative semidefinite. Therefore, we consider the case in which the output function is $h(x) = \Gamma x$, with $\Gamma \geq 0$ being the *inner coupling matrix*, defining the subset of state variables through which the network nodes are coupled. Accordingly, we select the following adaptive function:

$$g(x_j - x_i) = (x_j - x_i)^T \Gamma (x_j - x_i). \tag{2.12}$$

With this choice, we can further extend the stability results as shown in [39].

Theorem 2.3 *Let us assume that $h(x) = \Gamma x$, with $\Gamma \geq 0$, and that f is a continuous and differentiable vector field which is also QUAD(Δ, ω) with $\Delta = H\Gamma$, where H is an arbitrary diagonal matrix. Then, the network in Eq. 2.6 asymptotically synchronizes under adaptive law in Eq. 2.12 and all the coupling gains converge to finite values, that is, Eq. 2.10 and Eq. 2.11 hold.*

---

[1] A function $F : I \to \mathbb{R}$ is positive definite if $F(x) > 0$, $\forall x \in I$, $x \neq 0$ and $F(0) = 0$. A function $f : \mathbb{R}_{\geq 0} \to \mathbb{R}_{\geq 0}$ is of *class k* if it is continuous, positive definite and strictly increasing.

Fig. 2.1 Evolving networks. Bistable potential driving the evolution of each σ_{ij}

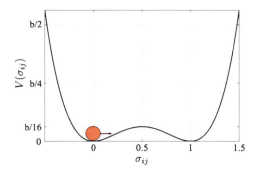

2.3.1 Evolving the Network Structure

In the adaptive scheme given in Eqs. 2.6, 2.7, each pair of nodes in the network is allowed to mutually negotiate the intensity of the coupling between themselves. Nonetheless, if at time 0 edge (i, j) is absent from the network, nodes i and j remain uncoupled for all time. In what follows, we describe the edge-snapping coupling mechanism presented in [8], where neighboring nodes are also allowed to activate or deactivate the link connecting them in such a way as for the network structure itself to evolve in time. This approach can be a very useful tool when the topology of the connections needs to be rearranged after edges/nodes failures while preserving the synchronous behavior. Specifically, we propose to model the gain evolution as a second order dynamical system:

$$\ddot{\sigma}_{ij} + \rho \dot{\sigma}_{ij} + \frac{d}{d\sigma_{ij}} V(\sigma_{ij}) = g(x_i, x_j). \qquad (2.13)$$

Here, the coupling gain σ_{ij} can be viewed as a unitary mass in a double-well potential V subjected to an external force g, function of the states of the neighboring nodes i and j, while ρ represents the damping. We assume that g vanishes as the network nodes synchronize.

A possible choice of the potential V is the following:

$$V(\sigma_{ij}) = b\sigma_{ij}^2(\sigma_{ij} - 1)^2. \qquad (2.14)$$

The shape of the potential is represented in Fig. 2.1, where b is a constant parameter regulating the depth of the wells. As can be observed from the figure, if the external force g dependent upon the distance between node i and node j is bounded and asymptotically vanishes, then the gain evolution determined by Eq. 2.13 must converge to one of its stable equilibria, that is, 0 (link is off) or 1 (link is on). Therefore, if synchronization is achieved, a topology emerges at steady-state, where only a subset of the possible edges of the network are activated.

A representative application of this adaptive approach to synchronization of a neural network is reported in Sect. 2.5.2

2.4 Inducing Synchronization via Design

An alternative strategy to achieve synchronization is to make sure that the properties of the node dynamics and the coupling functions are such that synchronization emerges from the network behavior. A powerful concept to assess this possibility is contraction theory as introduced in Sect. 2.2. Specifically, contraction theory can be effectively used to study synchronization by using a set of properties and results on contracting systems which are summarized briefly in what follows (all proofs can be found in [21, 33, 35]).

The key idea is to use the concept of *partial* contraction discussed in [38]. Basically, given a smooth nonlinear n-dimensional system of the form $\dot{x} = f(x, x, t)$, assume that the auxiliary system $\dot{y} = f(y, x, t)$ is contracting with respect to y. If a particular solution of the auxiliary y-system verifies a smooth specific property, then all trajectories of the original x-system verify this property exponentially. The original system is said to be *partially contracting*.

Indeed, the virtual y-system has two particular solutions, namely $y(t) = x(t)$ for all $t \geq 0$ and the particular solution with the specific property. Since all trajectories of the y-system converge exponentially to a single trajectory, this implies that $x(t)$ verifies the specific property exponentially.

We will now use the above basic results of nonlinear contraction analysis to study synchronization of complex networks by showing, first, that it is possible to effectively prove that a system (or virtual system) is contracting by means of an appropriate algorithmic procedure.

2.4.1 An Algorithm for Proving Contraction

As mentioned above, the key step in the use of contraction theory to achieve synchronization of networks is to prove that a given system of interest is contracting. In general, this can be a cumbersome task but, as recently shown in [34], the use of matrix measures and norms induced by non-Euclidean vector norms, (such as μ_1, μ_∞, $\|\cdot\|_1$, $\|\cdot\|_\infty$), can be effectively exploited to obtain alternative conditions to check for contraction of a dynamical system of interest, both in continuous-time and discrete-time.

In particular, we show now that by means of these measures and norms, it is possible to obtain an algorithmic procedure for checking if a system is contracting or for imposing such property. Formally, the conditions required by the algorithm for a system to be contracting are sufficient conditions. This means that, if the conditions are satisfied, then the system is contracting, while the vice-versa is not true.

The outcome of the algorithm is to provide a set of conditions on the elements of the system Jacobian, J, (and hence on the dynamics of $f(\cdot, \cdot)$) that can be used to prove contraction.

Here, we present an outline of the proposed algorithm for both continuous-time and discrete-time systems. The proof of the results on which the algorithm is based can be found in [34].

2 Adaptation and Contraction Theory for the Synchronization

The first step of the algorithm is to differentiate the system of interest, in order to obtain the Jacobian matrix, $J := \frac{\partial f}{\partial x}$:

$$\begin{bmatrix} J_{1,1}(x,t) & J_{1,2}(x,t) & \ldots & J_{1,n}(x,t) \\ J_{2,1}(x,t) & J_{2,2}(x,t) & \ldots & J_{2,n}(x,t) \\ \ldots & \ldots & \ldots & \ldots \\ J_{n,1}(x,t) & J_{n,2}(x,t) & \ldots & J_{n,n}(x,t) \end{bmatrix} \quad (2.15)$$

which is, in general, state/time dependent.

The next step of the algorithm is then to construct a directed graph from the system Jacobian. To this aim, we first derive an adjacency matrix from J, say $\mathscr{A}$, using the following rules:

1. initialize $\mathscr{A}$ so that $\mathscr{A}(i,j) = 0$, $\forall i, j$;
2. for all $i \neq j$, set $\mathscr{A}(i,j) = \mathscr{A}(j,i) = 1$ if either $J_{i,j}(x,t) \neq 0$, or $J_{j,i}(x,t) \neq 0$.

Such a matrix describes an undirected graph (see e.g. [13]), say $\mathscr{G}(\mathscr{A})$. The second step in the algorithm is then to associate directions to the edges of $\mathscr{G}(\mathscr{A})$ to obtain a directed graph, say $\mathscr{G}_d(\mathscr{A})$. This is done by computing the quantity

$$\alpha_{i,j}(x,t) = \frac{|J_{i,j}(x,t)|}{|J_{i,i}(x,t)|}(n - n_{0i} - 1). \quad (2.16)$$

In the above expressions n_{0i} is the number of zero elements on the i-th row of $\mathscr{A}$. (Note that if $J_{i,i}(x,t) = 0$ for some i, then, before computing (2.16), the system parameters/structure must be engineered so that $J_{i,i}(x,t) \neq 0$, for all i.)

The directions of the edges of $\mathscr{G}_d(\mathscr{A})$ are then obtained using the following simple rule:

the edge between node i and node j is directed from i to j if the quantity $\alpha_{i,j}(x,t) < 1$ while it is directed from j to i if $\alpha_{i,j}(x,t) \geq 1$.

Note that the quantities $\alpha_{i,j}(x,t)$ will be in general time-dependent, therefore the graph directions might be time-varying.

Once the directed graph $\mathscr{G}_d(\mathscr{A})$ has been constructed, contraction is then guaranteed under the following conditions:

1. uniform negativity of all the diagonal elements of the Jacobian, i.e. $J_{i,i}(x,t) < 0$ for all i;
2. for all t, the directed graph $\mathscr{G}_d(\mathscr{A})$ does not contain loops of any length;
3. $\alpha_{ij}(x,t)\alpha_{ji}(x,t) < 1$.

Note that, when the above conditions are not satisfied, the algorithm can be used to impose contraction for the system of interest by:

1. using, if possible, a control input to impose the first condition of the algorithm for all the elements $J_{i,i}(x,t)$ that do not fulfill it;
2. *re-direct* (using an appropriate control input, or tuning system parameters) some edges of the graph $\mathscr{G}_d(\mathscr{A})$ in order to satisfy the *loopless* condition;

3. associate to each reverted edge (e.g., the edge between node i and node j) one of the following inequalities:

   - $\alpha_{i,j}(x,t) \geq 1$, if the edge is reverted from j to i;
   - $\alpha_{i,j}(x,t) < 1$, if the edge is reverted from i to j.

2.4.2 Remarks

- Notice that the procedure presented above is based on the use of $\mu_\infty(\Theta J \Theta^{-1})$ and $\|\Theta J \Theta^{-1}\|_\infty$ for proving contraction. Other matrix measures and norms can also be used. In particular, for the continuous-time case, it is easy to prove that, using $\mu_1(\Theta J \Theta^{-1})$, yields the same algorithmic procedure applied on J^T. If this is the case, the resulting algorithm will follow the same logical steps presented above, with the only difference being the expression of $\alpha_{i,j}(x,t)$:

$$\alpha_{i,j}(x,t) := \frac{|J_{j,i}(x,t)|(m - c_{0i} - 1)}{|J_{i,i}(x,t)|}, \qquad (2.17)$$

where c_{0i} denotes the number of zero elements of the i-th column of J.
- The procedure presented here can have a clear physical interpretation. This is the case, for example, of molecular systems, that is, systems composed by genes and proteins, where the state variables represent the concentrations of the species involved into the system. In this case, each term $\alpha_{i,j}(x,t)$ represent a *normalized production rate* between species i and species j and the resulting set of inequalities provided by the algorithm points towards a balance of some *flow-like* quantities in the system (see, e.g., [31] and [30]);
- The *robustness* of the conditions outlined above can be evaluated. Specifically, we consider two cases: (i) robustness with respect to additive white noise acting on the system; (ii) robustness with respect to parameter variations. In the former case, if the conditions of the algorithm are fulfilled by the dynamics of the noise-free system, then all of its trajectories globally converge to a unique trajectory, say $\tilde{x}(t)$. Since our conditions ensure contraction, we have that (see [29]) all trajectories of the noisy system globally converge towards a boundary layer around $\tilde{x}(t)$. In the latter case, instead, the results provided by the algorithm are robust with respect to all parameter variations that do not affect the topology of the graph $\mathcal{G}_d(\mathcal{A})$. Such properties can be particularly useful for control and synchronization of networks having non-identical nodes and for biochemical systems which are typically characterized by high noise intensity and parameter uncertainties;
- Note that our approach can also be used to check partial contraction of the virtual system describing a network of oscillators. In this case, the algorithm is applied on the Jacobian of such a system and can be used to analyze synchronization phenomena in networks of coupled dynamical systems as will be further illustrated in Sect. 2.6.

2.5 Adaptive Synchronization of FitzHugh–Nagumo Neurons

To illustrate the theoretical results presented above and show the viability of the techniques discussed so far for the study of synchronization in neural systems, we consider now networks of FitzHugh–Nagumo (FN) oscillators. Such oscillators were first presented in [11] and their model take the form:

$$\dot{v} = c\left(v + w - \frac{1}{3}v^3 + u(t)\right),$$
$$\dot{w} = -\frac{1}{c}(v - a + bw),$$
(2.18)

where:

- v is the membrane potential;
- w is a recovery variable;
- $u(t)$ is the magnitude of the stimulus current.

Note that, equivalently, the model can be rewritten in the form $\dot{x} = f(x,t)$ where $x = (v, w)^T$ and $f(x) = (c(v + w - \frac{1}{3}v^3 + u(t)), -\frac{1}{c}(v - a + bw))^T$.

2.5.1 Properties of the Node Dynamics

We show here that networks of FitzHugh–Nagumo neurons can indeed be synchronized using a decentralized adaptive approach. As explained in Sect. 2.3, this is possible if the node dynamics satisfies the QUAD condition.

Computing the Jacobian J of the oscillator model Eq. 2.18, we have:

$$J(v) = \begin{bmatrix} c - v^2 & c \\ -\frac{1}{c} & -\frac{b}{c} \end{bmatrix}.$$
(2.19)

The symmetric part J_{sym} of J is:

$$J_{\text{sym}}(v) = \begin{bmatrix} c - v^2 & (c - \frac{1}{c})/2 \\ (c - \frac{1}{c})/2 & -\frac{b}{c} \end{bmatrix}.$$
(2.20)

From trivial matrix algebra, we have that

$$\lambda_{\max}(J_{\text{sym}}(v)) \leq \lambda_{\max}(J_{\text{sym}}(0)),$$
(2.21)

where

$$\lambda_{\max}(J_{\text{sym}}(0)) = \frac{-b + c^2 + \sqrt{1 + b^2 + 2(b - 1)c^2 + 2c^4}}{2c} := \gamma.$$
(2.22)

Also, we can write:
$$J_{sym}(v) = J^1_{sym} + J^2_{sym}(v), \quad (2.23)$$
where
$$J^1_{sym} = \begin{bmatrix} -d & (c - \frac{1}{c})/2 \\ (c - \frac{1}{c})/2 & -\frac{b}{c} \end{bmatrix}, \quad (2.24)$$
$$J^2_{sym}(v) = \begin{bmatrix} d + c - v^2 & 0 \\ 0 & 0 \end{bmatrix} \quad (2.25)$$

and d is some arbitrary positive scalar.

Now, we are ready to state the following results.

Theorem 2.4 *The FitzHugh–Nagumo oscillator is* QUAD(Δ, ω), *with* $\Delta - \omega I \geq \gamma I$.

Proof Assuming that $u(t)$ is continuously differentiable, also $f(x, t)$ will be continuously differentiable. Therefore, following [9], there exists a $\tilde{\xi} \in [0, 1]$ such that
$$(x - y)^T [f(x, t) - f(y, t)] = (x - y)^T \left(\frac{\partial}{\partial x} f(y + \tilde{\xi}(x - y), t) \right) (x - y). \quad (2.26)$$

Thus, from Eqs. 2.21–2.22, we have:
$$(x - y)^T [f(x, t) - f(y, t)] = \gamma (x - y)^T (x - y). \quad (2.27)$$

Comparing Eq. 2.27 with Eq. 2.3, the thesis follows. □

Theorem 2.5 *Let us assume that* $\Delta = H\Gamma$, *where H is an arbitrary diagonal matrix and*
$$\Gamma = \begin{bmatrix} 1 & 0 \\ 0 & 0 \end{bmatrix}.$$
The FitzHugh–Nagumo oscillator is QUAD(Δ, ω) *with* $H_{11} > d + c$.

Proof Considering that the Jacobian depends only on the first state variable v, Eq. 2.26 can be equivalently rewritten as
$$(x - y)^T [f(x, t) - f(y, t)] = (x - y)^T J_{sym}(\tilde{v})(x - y), \quad (2.28)$$
where $\tilde{v}$ is the first component of the point $\tilde{x} = y + \xi(x - y)$. From Eq. 2.23, we have
$$(x - y)^T [f(x, t) - f(y, t)] = (x - y)^T J^1_{sym}(x - y) + (x - y)^T J^2_{sym}(\tilde{v})(x - y). \quad (2.29)$$

It is easy to show that, for any c, we can choose a $d > 0$ such that $\lambda_{\max}(J^1_{\text{sym}}) = \mu$. Now, we can write

$$(x - y)^T [f(x,t) - f(y,t)]$$
$$\leq \mu(x - y)^T (x - y) + (d + c - \tilde{v}^2)(v_x - v_y)^2 \quad (2.30)$$
$$\leq \mu(x - y)^T (x - y) + (d + c)(v_x - v_y)^2. \quad (2.31)$$

Equation 2.31 can be rewritten as follows:

$$(x - y)^T [f(x,t) - f(y,t)] - (x - y)^T \begin{bmatrix} d+c & 0 \\ 0 & 0 \end{bmatrix} (x - y)$$
$$\leq \mu(x - y)^T (x - y). \quad (2.32)$$

Comparing Eq. 2.32 with Eq. 2.3, and considering that $\Delta = H\Gamma$ proves the theorem. $\square$

2.5.2 Numerical Validation

In what follows, we consider a network of FitzHugh–Nagumo coupled through the adaptive scheme in Eqs. 2.6–2.7. In particular, we assume that the oscillators are coupled linearly on the first state variable v. Namely, $h(x) = \Gamma x$, with

$$\Gamma = \begin{bmatrix} 1 & 0 \\ 0 & 0 \end{bmatrix}.$$

Moreover, according with Eq. 2.12, we choose:

$$\dot{\sigma}_{ij} = (v_i - v_j)^2, \quad (2.33)$$

so that neighboring oscillators adapt the strength of their coupling according to the distance between their membrane potentials.

It is easy to show that under these assumptions all the hypothesis of Theorem 2.3 are satisfied. In fact, from Theorem 2.5, the FitzHugh–Nagumo oscillator is QUAD with $\Delta = H\Gamma$.

As a numerical example, we consider a small world network of 500 nodes, with average degree equal to 8, and initial conditions taken randomly from a normal distribution.

As illustrated in Fig. 2.2, all the nodes' trajectories converge towards each other as expected with all coupling gains settling onto finite steady-state values.

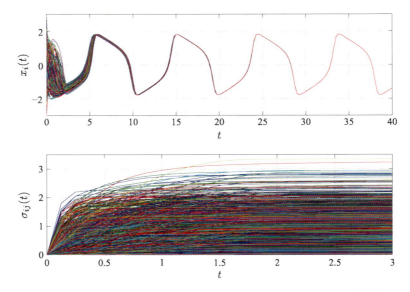

Fig. 2.2 Network of 500 FitzHugh–Nagumo oscillators coupled through the adaptive strategy Eq. 2.33. Evolution of the states (*top*) and of the adaptive gains (*bottom*)

2.5.2.1 Evolving the Network Topology

We consider now the case in which nodes can also decide whether to switch on or off a link between themselves according to the edge-snapping dynamics given by Eqs. 2.13, 2.14 as described in Sect. 2.3.1.

Specifically, we choose the external force g driving the gains' evolution as:

$$g(x_i, x_j) = (v_j - v_i)^2.$$

Moreover, we set $\sigma_{ij}(0) = 0$ so that, initially, all nodes are practically disconnected from each other. As we can see from Fig. 2.3, the gains then evolve converging towards 0 or 1 asymptotically so that a specific network topology emerges that guarantees synchronization.

2.6 Using Contraction

Now, we check if FitzHugh–Nagumo neurons can be synchronized via two alternative coupling strategies, diffusive coupling or local field potentials. In both cases, we will make use of contraction theory to assess the synchronizability of the network of interest and give conditions on the features of the node dynamics, the coupling and the network structure guaranteeing the emergence of a common asymptotic solution.

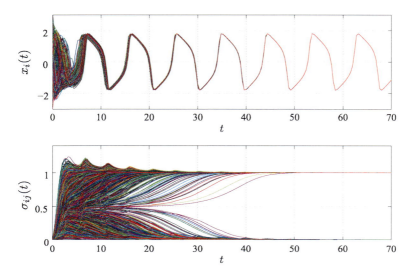

Fig. 2.3 Network of 500 FitzHugh–Nagumo oscillators coupled through the edge snapping strategy Eq. 2.13. Evolution of the states (*top*) and of the adaptive gains (*bottom*)

2.6.1 Node Dynamics

For the purpose of our analysis, it is convenient to define a new state variable, $w^* := \varepsilon w$, where ε is an arbitrary real number, and then write the system in the new coordinates (v, w^*). We have:

$$\begin{aligned}\dot{v} &= c\left(v + \frac{w^*}{\varepsilon} - \frac{1}{3}v^3 + u(t)\right), \\ \dot{w}^* &= -\frac{\varepsilon}{c}\left(v - a + b\frac{w^*}{\varepsilon}\right).\end{aligned} \quad (2.34)$$

Differentiation of the above system yields the Jacobian matrix

$$J := \begin{bmatrix} c - v^2 & \frac{c}{\varepsilon} \\ -\frac{\varepsilon}{c} & -\frac{b}{c} \end{bmatrix}. \quad (2.35)$$

Notice that the element $J_{11}(x, t)$ is not uniformly negative definite as required by the algorithm presented in Sect. 2.4.1. Hence, the node dynamics has to be somehow engineered in order to make the system contracting. We now propose a simple way to achieve this task. We will show later that this modification will be useful to derive sufficient conditions for network synchronization.

Specifically, we modify the system by adding an extra function $h(v)$ to the dynamics of v in (2.34). This is a simple form of feedback whose specific form will be selected so as to fulfill the condition $J_{ii}(x, t) < 0$. Differentiation of the modified dynamics yields:

$$J := \begin{bmatrix} c - v^2 + \frac{\partial h}{\partial v} & \frac{c}{\varepsilon} \\ -\frac{\varepsilon}{c} & -\frac{b}{c} \end{bmatrix}. \tag{2.36}$$

Therefore, in order to fulfill the condition $J_{11} < 0$, it immediately follows that a suitable choice for the function $h(v)$ is:

$$h(v) := -(c + \gamma_1)v \quad \text{for some positive scalar } \gamma_1.$$

Now, the Jacobian of the modified system is:

$$J := \begin{bmatrix} -\gamma_1 - v^2 & \frac{c}{\varepsilon} \\ -\frac{\varepsilon}{c} & -\frac{b}{c} \end{bmatrix}. \tag{2.37}$$

Thus, we have that both $J_{11}(x, t)$ and $J_{22}(x, t)$ are uniformly negative. Then, in order for the system to be contracting, we need to find conditions ensuring that no loops are present in $G_d(A)$. Since FN has a two dimensional phase space, it follows that $G_d(A)$ contains only two nodes. Thus, no loops are present in such a graph if:

$$\alpha_{12}(x, t) < 1;$$
$$\alpha_{12}(x, t) > 1,$$

or viceversa, with α_{ij} defined as in Eq. 2.16.

In our case, we have:

$$\alpha_{12} = \frac{|J_{12}(x, t)|}{|J_{11}(x, t)|} = \frac{c}{\varepsilon(\gamma_1 + v^2)},$$
$$\alpha_{21} = \frac{|J_{21}(x, t)|}{|J_{22}(x, t)|} = \frac{\varepsilon}{b}.$$

Now,

$$\alpha_{12} \leq \frac{c}{\varepsilon \gamma_1}.$$

Thus, the graph conditions are satisfied if:

$$\frac{c}{\varepsilon \gamma_1} < 1,$$
$$\frac{\varepsilon}{b} \geq 1.$$

Recall, now, that ε is an arbitrary real number. It then suffices to choose $\varepsilon > \max\{\frac{c}{\gamma_1}, b\}$ so that the above two conditions are satisfied. Finally, the last condition of the algorithm requires that

$$\alpha_{12}\alpha_{21} < 1.$$

Here, we have:

$$\alpha_{12}\alpha_{21} = \frac{c}{b\gamma_1}.$$

Thus, this condition is satisfied if:

$$\gamma_1 > \frac{c}{b}. \tag{2.38}$$

Therefore, an appropriate reengineering of the node dynamics and the use of the feedback input $h(v)$ can effectively make the FitzHugh–Nagumo oscillators contracting. In what follows, we will refer to such an input as a *contracting input* in accordance with what reported in [12].

Heuristically, from a physical viewpoint, the constraint represented by Eq. 2.38 can be interpreted as a lower bound on the strength of the feedback action introduced by the contracting input $h(v)$ on the node dynamics. This might be the principle behind the input to neurons in the brain provided by the small cortical circuit described in, for example, [4, 12], which may act as a frequency selector for many cortical pyramidal cells.

2.6.2 Synchronization Results

Now, we use the results obtained by means of the algorithm presented above to synchronize networks of N FN oscillators. We recall that the state variable of the i-th neuron is x_i, with $x_i := (v_i, w_i)^T$, while in this case we denote with $f(x_i)$ the dynamics of the i-th FN with the contracting input, that is,

$$f(x_i) := c\left(v + w - \frac{1}{3}v^3 + u(t)\right) + h(v_i) - \frac{1}{c}(v - a + bw),$$

where $h(v_i)$ is chosen as:

$$h(v_i) = -\delta(t)(c + \gamma_1)v_i$$

with $\delta(t) : \mathbb{R}^+ \to [0, 1]$ being some functions that determines when the contracting input is activated and $u(t)$ some periodic signal.

In what follows, we consider three different mechanisms for interconnecting different neurons. Specifically:

1. FN diffusively coupled;
2. FN coupled by means of field potentials;
3. FN with excitatory-only coupling.

2.6.2.1 Diffusive Coupling

In what follows, we will couple pairs of neighboring nodes only by means of the state variable v_i. Furthermore, we assume that the coupling function is linear. Specifically, let ϕ be any (coupling) function having diagonal Jacobian, with bounded nonnegative diagonal elements. The network considered here is then:

$$\dot{x}_i = f(x_i) + \sum_{j \in \mathscr{E}_i} \big(\phi(x_j) - \phi(x_i)\big), \tag{2.39}$$

where:

$$\phi(x) = \begin{pmatrix} Kx \\ 0 \end{pmatrix}.$$

In what follows, we make use of the following theorem, proved in [32], which generalize a previous result in [28].

Theorem 2.6 *Assume that in Eq. 2.39 the nodes dynamics are contracting. Then, all nodes dynamics globally exponentially converge towards each other, that is*:

$$|x_j(t) - x_i(t)| \to 0 \quad as\ t \to +\infty.$$

Furthermore, the rate of convergence towards the synchronous solution is the contraction rate of the network nodes.

In [32], an analogous of the above result is also given ensuring cluster synchronization. Clearly, such a choice for the coupling function satisfies the hypotheses required by Theorem 2.6.

Now, recall that the input $h(v)$ is a contracting input, with $\gamma_1 > c/b$ in the sense that it makes the system contracting. Thus, by means of Theorem 2.6, we have that when the nodes are diffusively coupled and the contracting input is activated, that is, $\delta(t) = 1$, then the nodes synchronize.

Figure 2.4 confirms this theoretical prediction. Specifically, we observe that all nodes start converging towards the same synchronous state from the moment the contracting input is activated at about $T \approx 2$.

2.6.2.2 Local Field Potentials

We now analyze the effects of a communication mechanism among neurons similar to the so-called quorum sensing in biological networks, for example, [32, 33]. In a neuronal context, a mechanism similar to that of quorum sensing may involve *local field potentials*, which may play an important role in the synchronization of clusters of neurons, [1, 10, 27, 37]. From a network dynamics viewpoint, the key characteristic of the quorum sensing-like mechanisms lies in the fact that communication between nodes (e.g., bacteria or neurons) occurs by means of a shared quantity. Furthermore, the production and degradation rates of such a quantity are affected by all

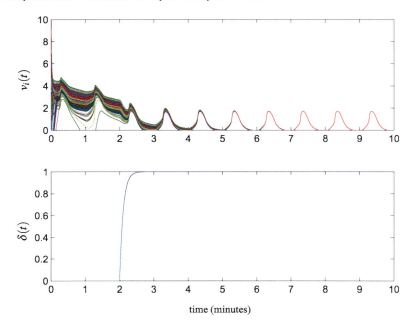

Fig. 2.4 Simulation of Eq. 2.39, with $N = 500$. Evolution of the oscillators' membrane potentials (*top panel*) and of the contracting input $\delta(t)$ (*bottom panel*). The system parameters are set as follows: $a = 0$, $b = 2$, $c = 2.5$, $K = 0.01$, $\gamma_1 = 1.5$ with $\delta(t) = [1 - e^{-10(t-2)}] \cdot H(t-2)$, $H(t)$ being the Heaviside function

the nodes of the network. Therefore, a detailed model of such a mechanism needs to keep track of the temporal evolution of the shared quantity, resulting in an additional set of ordinary differential equations:

$$\begin{aligned} \dot{x}_i &= f(x_i, z, t), \\ \dot{z} &= g(z, x_1, \ldots, x_N, t), \end{aligned} \qquad (2.40)$$

where $z(t) \in \mathbb{R}^d$ denotes the medium dynamics.

We need the following result from [32].

Theorem 2.7 *All nodes trajectories of the network in Eq. 2.40 globally exponentially converge towards each other if the function* $f(x, v_1(t), t)$ *is contracting for any* $v_1(t) \in \mathbb{R}^d$.

We will consider the case where the coupling between the nodes and the common medium is diffusive. In this case, the mathematical model in Eq. 2.40 reduces to:

$$\begin{aligned} \dot{x}_i &= f(x_i, t) + \phi_1(z) - \phi_2(x_i), \quad i = 1, \ldots, N, \\ \dot{z} &= g(z, t) + \sum_{i=1}^{N} \big[\phi_3(x_i) - \phi_4(z)\big]. \end{aligned} \qquad (2.41)$$

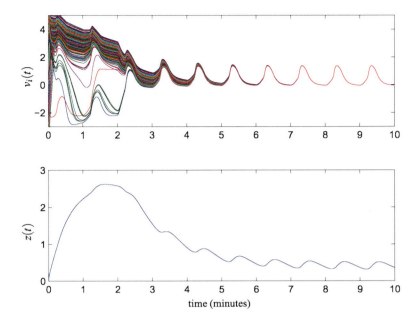

Fig. 2.5 Simulation of Eq. 2.41, with $N = 500$. System parameters are set as follows: $a = 0$, $b = 2$, $c = 2.5$, $K = 0.01$, $\gamma_1 = 1.5$

That is, the nodes and the common medium are coupled by means of the coupling functions $\phi_1 : \mathbb{R}^d \to \mathbb{R}^n$, $\phi_2 : \mathbb{R}^n \to \mathbb{R}^n$ and $\phi_3 : \mathbb{R}^n \to \mathbb{R}^d$, $\phi_4 : \mathbb{R}^d \to \mathbb{R}^d$. In our simulations, all the coupling functions are linear, that is,

$$\phi_1(z) = \begin{pmatrix} K \\ 0 \end{pmatrix} z, \qquad \phi_2(x_i) = \begin{pmatrix} K & 0 \\ 0 & 0 \end{pmatrix} \begin{pmatrix} v_i \\ w_i \end{pmatrix},$$

$$\phi_3(x_i) = \begin{pmatrix} K & 0 \end{pmatrix} \begin{pmatrix} v_i \\ w_i \end{pmatrix}, \qquad \phi_4(z) = Kz.$$

In this case, Theorem 2.7 implies that synchronization is attained and $z(t)$ remains bounded if $f(x, t) - \phi_2(x)$ is contracting.

Recall that we want the neurons to be weakly coupled. In particular, if K is sufficiently small this part of the input cannot make the system contracting. The behavior drastically changes when the contracting input is applied.

Figure 2.5 shows that when the contracting input is applied, a transition to the synchronous behavior occurs.

2.6.2.3 Excitatory-Only Coupling

We now turn our attention to the analysis of synchronization of excitatory-only coupled neurons, which are described by:

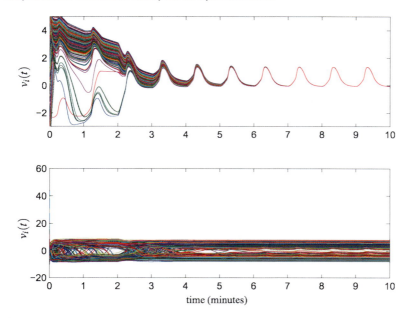

Fig. 2.6 Simulation of Eq. 2.41, with $N = 500$. System parameters are set as follows: $a = 0$, $b = 2$, $c = 2.5$, $\gamma_1 = 1.5$ $K = 0.01$ (*top*), $K = 5$ (*bottom*)

$$\dot{x}_i = f(x_i) + \sum_{j \in N_i} \phi(x_j), \qquad (2.42)$$

where the function ϕ is chosen as

$$\phi(x_j) := \begin{pmatrix} \gamma_2 & 0 \\ 0 & 0 \end{pmatrix} \begin{pmatrix} v_i \\ w_i \end{pmatrix}.$$

In [33], it has been proven that the above coupling function ensures synchronization if: (i) the contracting input is applied; (ii) γ_2 is *small* enough.

Figure 2.6 confirms the above prediction.

2.7 Materials & Methods

All the simulations have been performed using Matlab/Simulink, with variable step size ODE solver *ode45* with relative tolerance set to *1e-5*. Simulations for diffusive and excitatory-only coupling involved a directed small world network of $N = 500$ nodes.

2.8 Conclusions

In this chapter, we presented two alternative approaches to achieve synchronization of networks of interacting dynamical agents, one based on an adaptive coupling strategy, the other on the use of an appropriate contracting input. In the former case, we showed that, under the hypothesis that the node dynamics satisfies the so-called QUAD condition, it is possible for a set of agents to negotiate the strength of their mutual coupling so as to make a synchronous evolution emerge. The results obtained were applied to a representative network of FitzHugh–Nagumo neurons showing that indeed all neurons oscillate in synchrony with the strength of their couplings converging toward bounded constant values.

The alternative approach we took was that of considering contraction theory as a tool to engineer networks that synchronize. This can be seen as an effective design tool in all those cases where synchronization needs to be induced or suppressed by the appropriate engineering of the node dynamics and/or coupling structure and topology. After presenting the concept of contraction and briefly discussing its relevance for synchronization, we discussed a graphical procedure to check if a system or network of interest is contracting. This was then used to study the case of networks of FitzHugh–Nagumo oscillators, showing that, under the action of an *ad hoc* contracting input, the network can indeed synchronize. Three alternative coupling strategies were tested to show that synchronization can indeed be achieved in the presence of diffusive coupling, local mean field potentials or excitatory-only coupling.

The results presented show that synchronization is indeed a possible emergent phenomenon in current models of neuronal networks being used in the literature. Similar derivations can be applied to other neuronal models without major changes. The challenge now is to understand the exact role played by synchronization phenomena in the brain and the mechanism through which they emerge and are maintained even in the presence of disturbances and noise. In our view, the flexibility and robustness of these phenomena in the brain is probably due to the presence of the two features that were discussed in this chapter: adaptation of the network structure and coupling in time and specific properties of the node dynamics and interaction topology.

References

1. Anastassiou CA, Montgomery SM, Barahona M, Buzsaki G, Koch C (2010) The effect of spatially inhomogeneous extracellular electric fields on neurons. J Neurosci 30:1925–1936
2. Angeli D, Sontag ED (1999) Forward completeness, unboundedness observability, and their Lyapunov characterizations. Syst Control Lett 38:209–217
3. Bard E, Mason O (2009) Canars, clusters and synchronization in a weakly coupled interneuron model. SIAM J Appl Dyn Syst 8:253–278
4. Buzsaki G (2006) Rhythms of the brain. Oxford University Press, New York
5. De Lellis P, di Bernardo M, Garofalo F (2009) Novel decentralized adaptive strategies for the synchronization of complex networks. Automatica 45(5):1312–1318

6. De Lellis P, di Bernardo M, Garofalo F (2008) Synchronization of complex networks through local adaptive coupling. Chaos 18:037110
7. De Lellis P, di Bernardo M, Sorrentino F, Tierno A (2008) Adaptive synchronization of complex networks. Int J Comput Math 85(8):1189–1218
8. De Lellis P, di Bernardo M, Garofalo F, Porfiri M (2010) Evolution of complex networks via edge snapping. IEEE Trans Circuits Syst I 57(8):2132–2143
9. De Lellis P, di Bernardo M, Russo G (2011) On QUAD, Lipschitz and contracting vector fields for consensus and synchronization of networks. IEEE Trans Circuits Syst I 58(3):576–583
10. El Boustani S, Marre O, Behuret P, Yger P, Bal T, Destexhe A, Fregnac Y (2009) Network-state modulation of power-law frequency-scaling in visual cortical neurons. PLoS Comput Biol 5:e1000519
11. FitzHugh R (1961) Impulses and physiological states in theoretical models of nerve membrane. Biophys J 1:445–466
12. Gerard L, Slotine JJ Neural networks and controlled symmetries, a generic framework. Available at: http://arxiv1.library.cornell.edu/abs/q-bio/0612049v2
13. Godsil C, Royle G (2001) Algebraic graph theory. Springer, New York
14. Gonze D, Bernard S, Walterman C, Kramer A, Herzerl H (2005) Spontaneous synchronization of coupled circadian oscillators. Biophys J 89:120–129
15. Hartman P (1961) On stability in the large for systems of ordinary differential equations. Can J Math 13:480–492
16. Henson MA (2004) Modeling synchronization of yeast respiratory oscillations. J Theor Biol 231:443–458
17. Hong D, Sidel WM, Man S, Martin JV (2007) Extracellular noise-induced stochastic synchronization in heterogeneous quorum sensing network. J Theor Biol 245:726–736
18. Izhikevich EM (2006) Dynamical systems in neuroscience: the geometry of excitability and bursting. MIT Press, Cambridge
19. Lai CW, Chen CK, Liao TL, Yan JJ (2007) Adaptive synchronization for nonlinear FitzHugh–Nagumo neurons in external electrical stimulation. Int J Adapt Control Signal Process 22:833–844
20. Lewis DC (1949) Metric properties of differential equations. Am J Math 71:294–312
21. Lohmiller W, Slotine JJE (1998) On contraction analysis for non-linear systems. Automatica 34:683–696
22. McMillen D, Kopell N, Hasty J, Collins JJ (2002) Synchronization of genetic relaxation oscillators by intercell signaling. Proc Natl Acad Sci USA 99:679–684
23. Medvedev GS, Kopell N (2001) Synchronization and transient dynamics in the chains of electrically coupled FitzHugh–Nagumo oscillators. SIAM J Appl Math 61:1762–1801
24. Michel AN, Liu D, Hou L (2007) Stability of dynamical systems: continuous, discontinuous, and discrete systems. Springer, New York
25. Neiman A, Schimansky-Geier L, Cornell-Bell A, Moss F (1999) Noise-enhanced phase synchronization in excitable media. Phys Rev Lett 83(23):4896–4899
26. Pavlov A, Pogromvsky A, van de Wouv N, Nijmeijer H (2004) Convergent dynamics, a tribute to Boris Pavlovich Demidovich. Syst Control Lett 52:257–261
27. Pesaran B, Pezaris JS, Sahani M, Mitra PP, Andersen RA (2002) Temporal structure in neuronal activity during working memory in macaque parietal cortex. Nature 5:805–811
28. Pham QC, Slotine JJE (2007) Stable concurrent synchronization in dynamic system networks. Neural Netw 20:62–77
29. Pham QC, Tabareau N, Slotine JJE (2009) A contraction theory approach to stochastic incremental stability. IEEE Trans Autom Control 54:816–820
30. Russo G, di Bernardo M (2009) An algorithm for the construction of synthetic self synchronizing biological circuits. In: IEEE international symposium on circuits and systems, pp 305–308
31. Russo G, di Bernardo M (2009) How to synchronize biological clocks. J Comput Biol 16:379–393
32. Russo G, Slotine JJE (2010) Global convergence of quorum-sensing networks. Phys Rev E 82(4), submitted

33. Russo G, di Bernardo M, Sontag ED (2010) Global entrainment of transcriptional systems to periodic inputs. PLoS Comput Biol 6(4):e1000739
34. Russo G, di Bernardo M, Slotine JJE (2011) A graphical approach to prove contraction of nonlinear circuits and systems. IEEE Trans Circuits and Syst I 58(2):336–348
35. Slotine JJ (2003) Modular stability tools for distributed computation and control. Int J Adapt Control Signal Process 17:397–416
36. Sontag ED (1998) Mathematical control theory. Deterministic finite-dimensional systems. Springer, New York
37. Tabareau N, Slotine JJ, Pham QC (2010) How synchronization protects from noise. PLoS Comput Biol 6:e1000637
38. Wang W, Slotine JJE (2005) On partial contraction analysis for coupled nonlinear oscillators. Biol Cybern 92:38–53
39. Yu W, De Lellis P, Chen G, di Bernardo M, Kurths J (2010) Distributed adaptive control of synchronization in complex networks. IEEE Trans Autom Control, submitted
40. Zhou C, Kurths J (2006) Dynamical weights and enhanced synchronization in adaptive complex networks. Phys Rev Lett 96:164102

Chapter 3
Temporal Coding Is Not Only About Cooperation—It Is Also About Competition

Thomas Burwick

Abstract Temporal coding is commonly understood as a cooperative mechanism, realizing a grouping of neural units into assemblies based on temporal correlation. However, restricting the view on temporal coding to this perspective may miss an essential aspect. For example, with only synchronizing couplings present, the well-known superposition catastrophe may arise through states of global synchrony. Therefore, also desynchronizing tendencies have to be included, preferably in a manner that introduces an appropriate form of competition. In the context of phase model oscillator networks with Hebbian couplings, already a study of a "second-simplest" choice of model reveals a surprising property: temporal coding introduces a "competition for coherence" among stored patterns without need to include inhibitory couplings. Here, we refer to the corresponding models as Patterned Coherence Models (PCMs). This denotation is chosen with regard to the tendency of such models to establish coherence of parts of the network that correspond to the stored patterns, a valuable property with respect to pattern recognition. We review and discuss recent progress in studying such models, concentrating on models that complement synchronization with so-called acceleration. The latter mechanism implies an increased phase velocity of neural units in case of stronger and/or more coherent input from the connected units. Assuming Hebbian storage of patterns, it is demonstrated that the inclusion of acceleration introduces a competition for coherence that has a profound and favorable effect on pattern recognition. Outlook is given on paths towards including inhibition, hierarchical processing, and zero-lag synchronization despite time delays. We also mention possible relations to recent neurophysiological experiments that study the recoding of excitatory drive into phase shifts.

3.1 Introduction

Temporal coding is commonly understood as a particularly flexible mechanism for cooperation of neural units. Based on their temporal correlation, units of a network

---

T. Burwick (✉)
Frankfurt Institute for Advanced Studies (FIAS), Goethe-Universität, Ruth-Moufang-Str. 1, 60438 Frankfurt am Main, Germany
e-mail: burwick@fias.uni-frankfurt.de

"cooperate" (for example, they synchronize), resulting in a grouping of these units into an assembly such that the binding problem is resolved; see the reviews and discussions in the special issue of *Neuron* [1]. Notice, however, this perspective is missing a crucial piece. With only cooperating, for example, synchronizing couplings present, the complete network may easily run into states of global synchrony, where each unit carries the same phase. In consequence, the advantage of phases as markers of memberships to certain assemblies would be lost. This would reintroduce the well-known superposition catastrophe that motivated the proposal of temporal coding; see the review given by von der Malsburg [2]. Clearly, also competing tendencies have to be present to avoid the central problem that temporal coding is intended to solve.

Competition is then usually understood as "competition for activity", based on the inclusion of mutually inhibitory couplings. In this chapter, we argue that temporal coding provides, in the context of Hebbian storage of patterns, a different form of competition. This form of competition is referred to as "competition for coherence". In this contribution, we review recent progress related to the topic and emphasize the profound effect that competition for coherence may have on pattern recognition. Thereby, we want to promote the opinion that temporal coding is not only about cooperation between neural units—providing a binding into assemblies—but also about competition among patterns—providing the separation of the assembly from other activities.

Cooperation and Competition are basic mechanism underlying the dynamics of several complex systems. A concise exposition of this fact may be found in [3], with examples from physics, electrical engineering, economy, politics, epistemology, history and others. Thus, it should not be surprising if also models of brain dynamics and, accordingly, some brain-inspired computational intelligence may require an appropriate combination of cooperative and competitive mechanisms. The new aspect, reviewed and discussed here, is that temporal coding itself provides a special mechanism for competition, with properties not present in classical (that is, phase-independent) systems. Neglecting these properties could mean neglecting essential information processing benefits of temporal coding.

Any external input to a neural system, together with a given internal state, may excite a number of possible assemblies. Therefore, a competition among these assemblies is needed to arrive at unambiguous assembly formations. This unambiguous formation should result in coherent states of the winning assemblies (as proposed by the temporal correlation hypothesis), in contrast to decoherent or deactivated states of the competing and losing assemblies.

Recently, it was emphasized that such a competition for coherence may arise in the context of temporal coding in a surprisingly natural and simple manner. It may be based on excitatory couplings alone, without need to include inhibition (comments on a possible role of inhibition are given in Sect. 3.8).

The competition for coherence may result from properties that are obtained from standard phase model oscillatory systems with Hebbian memory by complementing synchronization with a so-called acceleration mechanism, an approach that we refer to as Patterned Coherence Model in the following (PCM, a denotation that is chosen

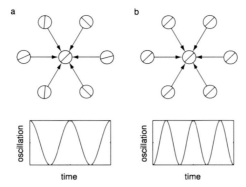

Fig. 3.1 This figures gives an illustration of the acceleration mechanism. The *upper figures* show some unit (displayed as the *central circle*) with connected units (displayed as the *outside circles*), where the *arrows* represent the connections to the central unit. The *strokes inside the circles* represent phases of these units. The *lower diagrams* illustrate the phase velocity of the central unit. The figure distinguishes between (**a**) a case where the incoming units have decoherent phases, while (**b**) shows a case of coherent incoming phases. The *lower figures* illustrate the effect of acceleration, that is, in case of (**b**), the phase velocity of the receiving unit is higher. As an essential ingredient of the discussed models, the increase in phase velocity is larger with stronger couplings (this effect is not illustrated). In case of (**a**), the phase velocity is given by intrinsic frequency and shear only, it is not "accelerated" through coherent signals from the connected units. (Notice, the figure is only a sketch of what is happening in a network: firstly, the *upper panels* show snapshots, while the *lower* show time periods; secondly, the recurrent couplings have to be taken into account to determine the dynamics of the complete network; see the discussion of the model in Sect. 3.6 and examples in Sect. 3.7)

with regard to the tendency of such models to establish coherence of parts of the network that correspond to the stored patterns; see Sects. 3.5 to 3.7).

The notion of acceleration refers to an increased phase velocity of a neural unit in case of stronger and/or more coherent input from the connected units [4–8] (see also [9] for a related but different foregoing approach). It is argued that acceleration arises naturally in the context of neural systems, both from an neural interpretation point of view, as well as from a formal perspective. See Fig. 3.1 for an illustration of the acceleration mechanism. In fact, the inclusion of acceleration is not a complicate, tricky, or artificial ingredient. On the contrary, it turns out to arise with already the "second-simplest" choice of phase coupling models (see Sect. 3.3 below). Nevertheless, studying its effect started only recently and answering related open questions may be obligatory for reaching a conclusion about the potential of temporal coding.

As a simple realization of a PCM, we study the Cohen–Grossberg–Hopfield (CGH) model [10, 11], generalized through combining it with phase dynamics. Assemblies are then identified with coherent sets of phase-locked patterns that are stored according to Hebbian memory. In the following, this perspective on assembly formation is reviewed and possible future directions of research are sketched.

In Sect. 3.2, some basic notation are introduced. Section 3.3 describes the notion of PCMs with respect to phase dynamic models. In Sect. 3.4, we refer to earlier work on phase model oscillatory networks. In Sect. 3.5, an argument is given for the benefit that may be obtained by using temporal coding. The argument will then lead to the generalized CGH model as PCM that is studied in Sect. 3.6. In Sect. 3.7, the effect on pattern recognition and temporal assembly formation is demonstrated with examples. In Sect. 3.8, an outlook is given on possible future research directions, in particular with respect to recent work described in [12, 13]. Section 3.9 contains the summary.

3.2 Basic Notations

Consider a network with N units, referred to through the indices $k, l = 1, \ldots, N$. The dynamical state of each unit k is described with an amplitude V_k and a phase θ_k. The amplitude is parameterized through a real parameter u_k so that

$$0 < V_k = g(u_k) = (1 + \tanh(u_k))/2 < 1. \tag{3.1}$$

Thus, g is the typical sigmoid activation function. The V_k may be referred to as activity of the unit k, where $V_k \simeq 1$ and $V_k \simeq 0$ refer to on- and off-states, respectively. In order to have also some temporal structure of the signal, we use the phase θ_k.

We assume that P patterns ξ_k^p, $p = 1, \ldots, P$, are stored and associated via so-called correlation matrix memory, a form of generalized Hebbian memory, given by the couplings

$$h_{kl} = \sum_{p,q,=1}^{P} d_{pq} \xi_k^p \xi_l^q \geq 0, \tag{3.2}$$

where for each k, p either $\xi_k^p = 0$ ("off-state") or $\xi_k^p = 1$ ("on-state"). The element d_{pq} may be thought of as implementing an association between two patterns p and q, described, for example, in [14, Sect. 2.11]. The standard Hebbian memory is given by the special case where the $P \times P$ matrix (d_{pq}) is diagonal (see Eq. 3.8 below).

Two patterns p and q may excite each other through the association given by d_{pq} or through common units. The latter may be described with the overlap matrix $\mathcal{O}$ given by

$$\mathcal{O}^{pq} = \sum_{k=1}^{N} \xi_k^p \xi_k^q. \tag{3.3}$$

It is useful to introduce measures to describe the collective dynamics of the patterns, that is, we introduce pattern activities A_p, coherences C_p, and phases Ψ_p, through

$$A_p = \frac{1}{N_p} \sum_{k=1}^{N} \xi_k^p V_k, \tag{3.4}$$

$$Z_p = C_p \exp(i\Psi_p) = \frac{1}{N_p A_p} \sum_{k=1}^{N} \xi_k^p V_k \exp(i\theta_k), \tag{3.5}$$

where N_p is the number of units k with $\xi_k^p = 1$ [9]. Notice, $0 \leq C_p \leq A_p \lesssim 1$ and C_p takes the maximal value, $C_p \simeq 1$, if the active units of pattern p, that is, the units k' that are not off-state, have the same phase, $\theta_{k'} = \Psi_p$.

3.3 Patterned Coherence Models (PCMs)

In Sects. 3.5 to 3.7, we consider oscillatory neural network models that have amplitude and phase dynamics. These models have a simple structure of the phase dynamics. In this section, we give a separate description of it in order to emphasize the simplicity.

Consider the following phase dynamics:

$$\frac{d\theta_k}{dt} = \omega_{0,k} + \frac{\gamma}{\tau} \sum_{l=1}^{N} h_{kl} \sin(\theta_l - \theta_k + \chi), \quad (3.6)$$

where the h_{kl} are given by Eq. 3.2, t is time, τ is a time-scale, $\omega_{0,k}$ are intrinsic frequencies, and $\gamma > 0$. The case $\chi = 0$ has been studied extensively (see the next section for references). Therefore, it may come as a surprise that this equation could hold any unexpectedly useful behavior. The main reason for studying the case $\chi = 0$ was to avoid desynchronizing effects. Recently, however, it was emphasized that the desynchronizing tendencies that follow from $\chi \neq 0$, $\cos(\chi) > 0$, may have the effect of introducing a competition for coherence that, when combined with amplitude dynamics, has a profound and favorable effect on pattern recognition [4–6].

In the Introduction, Sect. 3.1, we mentioned that the characteristic properties discussed in this chapter arise with studying already the second-simplest choice of phase models. This statement referred to the fact that the Kuramoto model (with $\chi = 0$) is generally considered to be the simplest model for synchronization. Correspondingly, Kuramoto and his coworkers referred to the models with $\chi \neq 0$ as "second-simplest" models [15].

Actually, apart from the effect of introducing a competition for coherence (the property that is explained in Sects. 3.5 to 3.7), there are at least three reasons to allow for $\chi \neq 0$, $\cos(\chi) > 0$ (the restriction to $\cos(\chi) > 0$ is required to preserve synchronization, see Sect. 3.6.2 below).

Firstly, the case $\chi \neq 0$, $\cos(\chi) > 0$, has a natural dynamical interpretation in terms of neural dynamics. As explained in Sect. 3.6.2, it introduces the acceleration mechanism that was mentioned in Sect. 3.1 and is sketched in Fig. 3.1.

Secondly, the case $\chi \neq 0$ arises also naturally from a formal point of view. It was shown that the model discussed in Sects. 3.5 to 3.7 may be obtained as a complex-valued version of the classical Cohen–Grossberg–Hopfield (CGH) gradient system. Allowing also for nonvanishing imaginary parts of the coupling term coefficients implies the case $\chi \neq 0$ (the real parts imply synchronization, the imaginary parts imply acceleration) [4].

Thirdly, perhaps the most vague but potentially the most fascinating aspect (depending, of course, on personal taste): the phase shifts expressed through $\chi \neq 0$, resulting in the acceleration effect, may be linked to experimental observations of phase shifts that are found with synchronized states of brain dynamics. Part of their interpretation run under the notion of "recoding of excitatory drive into phase shifts"; see the additional remarks and references in Sect. 3.8.

In Sect. 3.6.2, a simple trigonometric identity is used to show that $\chi \neq 0$, $\cos(\chi) > 0$, introduces the acceleration mechanism that was mentioned in Sect. 3.1. Thus, the notion of a PCM corresponds to the presence of a nonvanishing phase χ in dynamical models with phase dynamics related to Eqs. 3.2 and 3.6.

3.4 Earlier Approaches Based on Phase Models

In this section, before going into the details of studying a neural network version of a PCM in Sects. 3.5 to 3.7, we briefly mention earlier approaches based on networks of phase model oscillators.

Following the proposal of the temporal correlation hypothesis by Christoph von der Malsburg in 1981 [16], see also [17, 18], it was realized that oscillatory networks are natural candidates for implementing temporal coding [19–23]. At that time, impressive new results on oscillatory networks based on phase model oscillators, obtained as effective models for complex-valued Ginzburg–Landau systems, were summarized by Kuramoto in [24]. (Notice, how closely many of the foregoing references are related to the field of synergetics, that is, the study of self-organizing complex systems, initiated by Hermann Haken around 1980.) A more recent review of the Kuramoto model may be found in [25]. Also due to the fact that networks of phase models oscillators allow for simple formulations, this direction of modeling oscillatory systems was followed by several attempts to model temporal coding.

As a natural step to bring the Kuramoto model closer to neural modeling, the all-to-all couplings of the Kuramoto model were replaced with Hebbian couplings of the kind described with Eq. 3.2 (actually, using only a diagonal form of d_{pq}, that is, standard Hebbian memory, not correlation matrix-memory). Around 1990, this led to studying the kind of models given by Eq. 3.6 with

$$\chi = 0. \tag{3.7}$$

This approach was soon extended to include also amplitude dynamics, using also different complex-valued formulations; see the lists of references given in [5, 9]. A summarizing review was given in [26]. Recent reviews of related approaches based on complex-valued networks may be found in [27, 28]. With a view on the introductory remarks given in Sect. 3.1, it may be said that these early approaches established some consistent framework for implementing synchronization as the cooperative mechanism.

As mentioned in Sect. 3.3, the choice $\chi = 0$ serves to avoid desynchonization. However, the original proposal of temporal coding, as given in [16] and reviewed in [2], was motivated by finding ways to avoid the superposition catastrophe that was observed by Rosenblatt already in the early 60s [29]. It should be noticed that establishing only synchronization without desynchronization of the parts of the network that do not participate in the assembly may imply global synchronization. This would reestablish the superposition catastrophe, making temporal coding useless (see Example 1 in Sect. 3.7.2). Thus, a complete formulation has to implement also desynchronizing mechanisms, preferably in a manner that establishes competition among possible assemblies. It is then a pleasant fact that the desynchronization that follows from $\chi \neq 0$, $\cos(\chi) > 0$, may already be sufficient to introduce a form of desynchronization that supports pattern recognition through complementing the cooperation with competition (as explained in Sects. 3.5 to 3.7).

Due to the simplicity of Eq. 3.6, related models have appeared several times before. However, as mention above, usually the phase χ was set to zero to avoid desynchronization. An early study of the case $\chi \neq 0$, with nearest neighbor couplings, that is, $h_{kl} = K$, for k and l nearest neighbors, and $K > 0$ some coupling strength, while $h_{kl} = 0$ otherwise, was given in [15]. It is interesting to read the authors' motivation:

> In living systems such as mammalian intestine and heart, the cells forming a tissue oscillate with frequencies larger than when they are isolated from each other [15],

giving also references to [30, 31]. Therefore, assuming acceleration amounts to assuming in a neural context what is known to be true for other parts of living systems. Some of the few other references to studies of models with $\chi \neq 0$ may be found in the review of the Kuramoto model in [25, Sect. IV.B].

Perhaps closest in spirit to the approach discussed in the following are some models of attention, where networks synchronize only partially, controlled by phase interaction with a central unit; see, for example, [32, 33] and references therein. However, these studies do not concern Hebbian memory, an essential ingredient of the following discussion. Moreover, partial synchronization in the following is not caused by including a central oscillator.

3.5 Why Temporal Coding? The Argument for Patterned Coherence Models

To make the following argument easier to understand, we restrict our discussion to the standard Hebbian memory case, that is, we assume that

$$(d_{pq}) = \mathrm{diag}(\lambda_1, \ldots, \lambda_P), \tag{3.8}$$

where each $\lambda_p > 0$ weights the contribution of pattern p to the couplings. (We return to commenting on more general d_{pq} only in Sect. 3.8.)

To value the benefits that temporal coding may provide, it is useful to remember a reformulation of classical couplings, that is, couplings in artificial neural networks where the signals are described in terms of activity alone, without including phase coordinates. For each unit k, its input from other units may be expressed in terms of pattern activities:

$$\sum_{l=1}^{N} h_{kl} V_l = \sum_{p=1}^{P} \lambda_p \xi_k^p \sum_{l=1}^{N} \xi_l^p V_l = \sum_{p=1}^{P} \xi_k^p \lambda_p N_p A_p, \qquad (3.9)$$

with pattern activities, $0 \lesssim A_p \lesssim 1$, as defined by Eq. 3.4. Notice, due to the factor ξ_p^k the unit k couples only to patterns that are on-state at unit k, that is, to patterns p with $\xi_k^p = 1$.

Let us make a corresponding reformulation for synchronizing terms like the ones appearing in Eq. 3.6. For the moment, we set $\chi = 0$. Moreover, we extend the couplings terms by including the activity V_k to assure a coupling to active units only. Then, in analogy to Eq. 3.9, one may easily obtain

$$\sum_{k=1}^{N} h_{kl} \sin(\theta_l - \theta_k) V_l$$

$$= \operatorname{Im} \sum_{p=1}^{P} \lambda_p \xi_k^p \exp(-i\theta_k) \sum_{l=1}^{N} \xi_l^p V_l \exp(i\theta_l)$$

$$= \sum_{p=1}^{P} \xi_k^p \lambda_p N_p A_p C_p \sin(\Psi_p - \theta_k). \qquad (3.10)$$

This equation is a remarkable identity. Although almost trivial to derive, it contains in a nutshell the essence of what temporal coding may provide for information processing. In particular, a comparison with Eq. 3.9 makes obvious what temporal coding provides beyond the classical (that is, phase-independent) coding. This point of view is motivated in the following.

Given the form of Eq. 3.10 as a sum of synchronizing terms, its form in terms of pattern quantities implies that each term in the sum introduces a tendency to synchronize unit k with the phase Ψ_p of each pattern p that couples to unit k, that is, each pattern with $\xi_k^p = 1$.

Something analogous is true already for the classical couplings as described by Eq. 3.9. There, each pattern p with $\xi_k^p = 1$ introduces a tendency to activate the unit k. The crucial difference is that in case of the classical coupling each pattern "pulls" the unit k into "the same direction". That is, each pattern introduces the same tendency to activate unit k, the patterns do not carry a marker that would distinguish them from each other. This is the shortage of the classical formulation pointing to the superposition catastrophe that motivated the temporal correlation hypothesis [2, 16].

The situation changes decisively if phases are taken into account. Then, with different pattern phases Ψ_p in Eq. 3.10, the terms given by $\sin(\Psi_p - \theta_k)$ may introduce different tendencies, they may "pull" unit k into "different directions" given by the different pattern phases. These provide markers that make the patterns distinguishable also from the perspective of a single unit k. This is why Eq. 3.10 points to the essential benefit that temporal coding provides.

There is, however, also another lesson that should be learned from this argument. In fact, it may be judged to provide the essential motivation to consider PCMs. The above argument crucially depends on the pattern phases taking different values. This hints at a necessity of implementing temporal coding through synchronization in combination with desynchronizing tendencies that let the pattern phases take different values. PCMs do just that, when appropriately embedded into a neural network framework, as explained in Sect. 3.6. There, the discussed model implements a tendency to let the pattern phases oscillate with different frequencies due to the resulting acceleration effect. The different frequencies of the pattern phases may then make global synchronization impossible. In consequence, a competition for coherence arises among the patterns, thereby separating patterns in a way that would not be possible if phase information is not included. The next section describes the argument with more details.

3.6 Oscillatory Neural Network Model with Synchronization and Acceleration

3.6.1 Simple Ansatz for Phase and Amplitude Dynamics

The following PCM may be derived from a gradient system that is a complex-valued generalization of the classical Cohen–Grossberg–Hopfield (CGH) system [10, 11]; see [4] for the derivation. In this section, however, we rather want to approach the model from a heuristic perspective.

Let us search for an amplitude dynamics that is close to a classical neural network model, the CGH model, while the phase dynamics should be a generalization of the model given by Eq. 3.6. The two requirements may then be realized with the following ansatz:

$$\tau \frac{du_k}{dt} = \Lambda(u_k)\left(-u_k + I_k + \sum_{l=1}^{N} c_{kl}(u,\theta) V_l\right), \quad (3.11a)$$

$$\frac{d\theta_k}{dt} = \omega_{0,k} + \omega_{1,k} V_k + \frac{1}{\tau}\sum_{l=1}^{N} s_{kl}(u,\theta) V_k, \quad (3.11b)$$

where I_n are external inputs and $\omega_{1,k}$ are shear parameters. The factor $\Lambda(u_k)$ allows for a dynamical time-scale of the amplitude dynamics. The amplitude V_k that is related to coordinate u_k by Eq. 3.1, the phases θ_k, the time scale τ and intrinsic frequencies $\omega_{0,k}$ were already introduced in Sect. 3.2.

As close-to-minimal choices for the couplings in Eqs. 3.11a, 3.11b we set:

$$c_{kl}(u,\theta) = h_{kl}(\alpha + \beta \cos(\theta_l - \theta_k + \phi)), \quad (3.12a)$$

$$s_{kl}(u,\theta) = h_{kl}\gamma \sin(\theta_l - \theta_k + \chi). \quad (3.12b)$$

(Couplings with higher-mode and amplitude dependencies were derived from the complex-valued gradient system in [4, Sect. 6]. Here, we restrict the discussion to the simple choices given by Eqs. 3.12a, 3.12b.) The $\alpha \geq 0$ describes the classical CGH couplings, while temporal coding enters through the trigonometric functions. Let us first assume $\phi = \chi = 0$. Then, sine and cosine implement synchronizing tendencies between the units. With $\gamma > 0$, the sine describes the tendency to synchronize θ_k towards θ_l. Correspondingly, with $\beta > 0$, the cosine describes a strengthening of the couplings in case of synchronization.

As mentioned in the foregoing sections, the choice of nonvanishing phase χ turns out to be crucial. This will be discussed beginning from Sect. 3.6.2. Notice also, the parameters β, ϕ may be related to γ, χ and the choice of Λ may be fixed through assuming the complex-valued formulation. More will be said on this in Sect. 3.6.2.

3.6.2 Synchronization and Acceleration

The dynamical content given by Eqs. 3.11a, 3.11b, 3.12a, 3.12b gets clearer when using the trigonometric identity

$$\gamma \sin(\theta_l - \theta_k + \chi) = \tau \omega_2 \cos(\theta_l - \theta_k) + \sigma \sin(\theta_l - \theta_k), \quad (3.13)$$

where

$$\tau \omega_2 = \gamma \sin(\chi) \quad \text{and} \quad \sigma = \gamma \cos(\chi). \quad (3.14)$$

Using this identity, we may reformulate the phase dynamics given by Eqs. 3.11b and 3.12b as

$$\frac{d\theta_k}{dt} = \widehat{\omega}_k(u, \theta) + \underbrace{\frac{\sigma}{\tau} \sum_{l=1}^{N} h_{kl} \sin(\theta_l - \theta_k) V_n}_{\text{synchronization terms}}, \quad (3.15)$$

where

$$\widehat{\omega}_k(u, \theta) = \omega_{0,k} + \omega_{1,k} V_k + \underbrace{\omega_2 \sum_{m=1}^{N} h_{kl} \cos(\theta_l - \theta_k) V_k}_{\text{acceleration terms}}. \quad (3.16)$$

The terms proportional to ω_2 are denoted as acceleration terms. It should be obvious from Eq. 3.16 that the acceleration mechanism implies for each unit k an increase in phase velocity for more coherent and/or stronger input from the connected units, as illustrated with Fig. 3.1; see Fig. 3.2 for an illustration of the different contributions to $\widehat{\omega}_k$, as well as the form of the acceleration terms given by Eqs. 3.18 and 3.19 and the related discussion in Sect. 3.6.3.

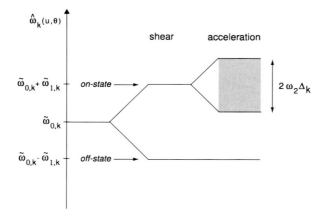

Fig. 3.2 The figure illustrates different parts of the phase velocity contribution $\widehat{\omega}_k(u,\theta)$ given by Eq. 3.16; see also Eqs. 3.18 and 3.19. Intrinsic frequencies and shear parameter were reparameterized such that $\omega_{0,k} + \omega_{1,k} V_k = \widetilde{\omega}_{0,k} + \widetilde{\omega}_{1,k} \tanh(u_k)$. Two cases are illustrated. If unit k is off-state and also the connected units are off-state, then $\omega_k(u,\theta)$ is given by $\widetilde{\omega}_{0,k} - \widetilde{\omega}_{1,k} = \omega_{0,k}$. If unit k is on-state, then $\widehat{\omega}_k(u,\theta)$ is increased due to nonvanishing shear parameter. Moreover, acceleration allows for a spectrum of possible values, indicated through the gray region (the quantity Δ_k is defined in Sect. 3.6.3). The actual value of the contribution from acceleration is inside of this spectrum; see the discussions in Sects. 3.6 and 3.7

In case of symmetric couplings h_{kl}, it is possible to derive Eqs. 3.11a, 3.11b and 3.12a, 3.12b from a complex-valued gradient system, provided that [5]

$$\gamma = 2\beta, \qquad \phi = \chi \quad \text{and} \quad \Lambda = (1 - V_n)^{-1}. \tag{3.17}$$

Here, we will not review this property. Nevertheless, with our examples in Sect. 3.7, we use the parameter dependencies given by Eqs. 3.17 (see also the remarks on alternative choices for $\Lambda(u_k)$ in [4, Sect. 6]).

3.6.3 Competition for Coherence

Let us abbreviate the sum of acceleration terms in Eq. 3.16 as Γ_k, that is,

$$\Gamma_k(u,\theta) = \widehat{\omega}_k(u,\theta) - \omega_{0,k} - \omega_{1,k} V_k. \tag{3.18}$$

Steps analogous to the ones that implied Eq. 3.10 allow to give them a form that makes their effect more obvious:

$$\Gamma_k(u,\theta) = \omega_2 \sum_{p=1}^{P} \xi_k^p \lambda_p N_p A_p C_p \cos(\Psi_p - \theta_k). \tag{3.19}$$

Notice, due to the presence of ξ_k^p in the reformulated sum, θ_k is affected only by patterns that couple to unit k, that is, patterns p with $\xi_k^p = 1$. In consequence, this

Fig. 3.3 A detailed view on the spectrum displayed in Fig. 3.2 (notice, here, the axis shows the acceleration terms Γ_k, while the axis in Fig. 3.2 showed $\widehat{\omega}_k = \omega_1 + \omega_2 V_k + \Gamma_k$). Hebbian memory combined with acceleration implies the pattern-frequency-correspondence and, thereby, the competition for coherence among the patterns and possible assemblies; see Sect. 3.6.3

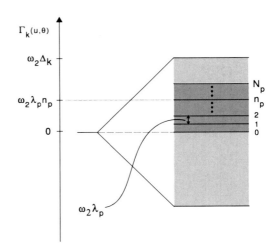

implies that acceleration introduces a desynchronization of the network, because different units are "accelerated" by different sets of patterns. The maximal acceleration at each unit k is given by $\Gamma_k(u, \theta) \simeq \omega_2 \sum_{p=1}^{P} \xi_k^p \lambda_p = \omega_2 \Delta_k$. (Assume complete activation and coherence of the patterns, implying $\Psi_p = \Psi$, and set $\theta_k = \Psi$ and $\theta_k = \Psi + \pi$.)

Consider a particular pattern p and assume that n_p of its units are on-state, while the remaining $N_p - n_p$ units are off-state. In terms of the pattern activity, this implies

$$0 \lesssim A_p \simeq \frac{n_p}{N_p} \leq 1. \tag{3.20}$$

Moreover, we consider the following idealized situation. Assume that the pattern is coherent, $C_p \simeq 1$, while the overlapping patterns $q \neq p$ are assumed to be decoherent, $C_q \simeq 0$. Then, the on-state units k of pattern p are in a state that we refer to as pure pattern state, where

$$\Gamma_k(u, \theta) \simeq \omega_2 \lambda_p n_p \quad \text{(pure pattern state)}. \tag{3.21}$$

In the context of our examples in Sect. 3.7, we want to demonstrate the effect of acceleration. To isolate the desynchronization resulting from acceleration and not to confuse it with desynchronization due to different values of intrinsic frequencies and shear parameters, we choose to use identical values for these. Thus, we set

$$\omega_{0,k} = \omega_0 \quad \text{and} \quad \omega_{1,k} = \omega_1, \tag{3.22}$$

for $k = 1, \ldots, N$, and some ω_0, ω_1. (See [6] for examples with nonidentical values.) Assuming then a pure pattern situation, we find that a pure state of pattern p would have phase velocity given by

$$\Omega_p^{n_p} = \omega_0 + \omega_1 + \omega_2 \lambda_p n_p \quad \text{(pure pattern frequency)}. \tag{3.23}$$

This frequency takes different values for different numbers n_p of on-state units of pattern p, implying a spectrum of possible pure pattern frequencies; see Fig. 3.3. We refer to Eqs. 3.21 and 3.23 as pattern-frequency-correspondence [5].

With a view on the pattern-frequency correspondence, one may expect that each pattern introduces a tendency to synchronize with a different frequency. Notice then, this is exactly the ingredient to temporal coding that was considered to be relevant at the end of Sect. 3.5. There, it was argued that such a property may imply a competition for coherence among the patterns and related assemblies. According to Eq. 3.10 with Eq. 3.20, a single pattern p may win this competition due to a larger value of $n_p \lambda_p$, that is, the number of on-state units times the pattern weight (but also a winning assembly with several patterns is possible, see Example 5). This behavior is demonstrated with the following examples.

3.7 Examples

3.7.1 Network Architecture and Parameter Choices

We study the PCM given by Eqs. 3.11a, 3.11b, 3.12a, 3.12b, and 3.17, and consider a network with $N = 30$ units and $P = 6$ stored patterns. The overlap between patterns p and q may be described through the overlap matrix defined in Eq. 3.3. The patterns of the following examples have

$$(\mathcal{O}^{pq}) = \begin{pmatrix} 15 & 4 & 1 & 3 & 1 & 2 \\ . & 10 & 0 & 2 & 1 & 0 \\ . & . & 5 & 0 & 1 & 0 \\ . & . & . & 4 & 0 & 0 \\ . & . & . & . & 4 & 0 \\ . & . & . & . & . & 4 \end{pmatrix}, \qquad (3.24)$$

where the symmetric elements are not repeated. For example, patterns $p = 2$ and $p = 4$ have $N_1 = 10$ and $N_4 = 4$ units, respectively, and they are overlapping at two units. We ordered the patterns so that $N_p \geq N_{p+1}$. With our examples, standard Hebbian memory is assumed as described with Eqs. 3.2 and 3.8. The default values of the pattern weights are given by $\lambda_p = 1/P$ for every p.

The default parameters are chosen to be $\alpha = 8$ and $\sigma = \tau \omega_2 = 2\pi$, $\tau = 1$. Moreover, we set $\omega_{0,k} = \omega_{1,k} = 0$. This allows to trace the effect of Eq. 3.18 back to the acceleration terms alone, without confusing it with effects of intrinsic or shear-induced frequencies. The inputs are chosen so that the network is completely activated.

In the following, only Example 2 uses exactly these default values. Each of the other example uses one or more parameters that differ from these default values in order to illustrate the effect of these parameters.

The discretization is realized through using an Euler approximation with discrete time step $dt = \epsilon \tau$, $0 < \epsilon \ll 1$, and replacing $\Lambda(u_k) \to \Lambda_\epsilon(u_k) = (1 - V_k + \epsilon)^{-1}$. We chose $\epsilon = 0.01$.

The initial values for amplitudes u_k are small random values, while the initial phases θ_k are randomly distributed in the interval from 0 to 2π. The examples use the same set of initial values.

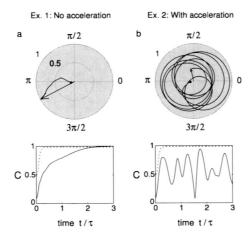

Fig. 3.4 (a) Example 1 is without acceleration, $\omega_2 = 0$, while (b) Example 2 uses acceleration, $\tau\omega_2 = \sigma > 0$. The *upper panels* show the dynamics of global coherence C and phase Ψ (defined in Sect. 3.7.2) through displaying the trajectory of $Z = C\exp(i\Psi)$ for $t = 0$ to $t = 3\tau$ on the complex-valued unit disk. The *dots* show the beginning of the trajectory at $t = 0$, while the *arrows* point to the end of the trajectory at $t = 3\tau$. The *lower diagrams* show the time course of the global coherence C (*solid line*) and activity A (*broken line*). The regularity of case (**b**) gets obvious when considering the dynamics in terms of pattern quantities as displayed in Fig. 3.5. See the discussion in Sect. 3.7.2

3.7.2 Examples 1 and 2: Effect of Including Acceleration

The first two examples illustrate the desynchronizing effect of acceleration and, related to this, its implication of introducing the competition for coherence.

To illustrate the desynchronizing effect, we introduce the notions of global activity, coherence and phase in analogy to the quantities given by Eqs. 3.4 and 3.5. There we replace the pattern vectors ξ_p^k with a vector $e_k = 1$ for each k, and A_p, Z_p, C_p, Ψ_p are replaced with A, Z, C, Ψ.

Example 2 uses the default parameter values specified in Sect. 3.7.1, in particular, a nonvanishing value of the acceleration parameter, $\tau\omega_2 = \sigma > 0$. In contrast, Example 1 uses vanishing acceleration, $\omega_2 = 0$. The global dynamics differs significantly, see Fig. 3.4. The input was chosen so that the network is completely activated, that is, each unit gets on-state. Example 1, the case without acceleration, $\omega_2 = 0$, shows global synchrony and therefore the presence of the superposition catastrophe. In contrast, Example 2, differing only by using acceleration, $\tau\omega_2 = \sigma > 0$, shows the desynchronizing effect of acceleration.

Looking at Fig. 3.4, one may be tempted to avoid acceleration (correspondingly, to set $\chi = 0$ in Eq. 3.6), since no regularity—and therefore usable property—appears recognizable in the dynamics. However, this would be a misled conclusion, as gets obvious when considering the dynamics with respect to pattern quantities; see Fig. 3.5.

3 Temporal Coding Is Not Only About Cooperation—It Is Also About Competition   47

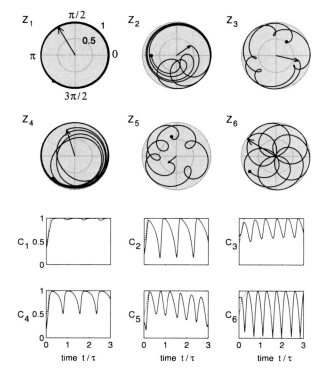

Fig. 3.5 Example 2. The competition for coherence arises as a consequence of combining synchronization with acceleration. Due to this competition, only the dominating pattern $p = 1$ takes a state of enduring coherence. The other patterns return to states of decoherence, indicated by the cycloidal trajectories of the Z_q for $q \neq 1$. The *panels at the top* show pattern coherences C_p and pattern phases Ψ_p through displaying the trajectories of $Z_p = C_p \exp(i\Psi_p)$ on the complex-valued unit disk. The *dots* show the beginning of the trajectory at $t = 0.5\tau$, while the *arrows* point to the end of the trajectory at $t = 3\tau$. The *panels at the bottom* show the time course of coherence C_p in a diagrammatic view, displayed by the *solid lines*. The *dotted lines* show the pattern activities A_p, confirming that inputs and initial values have been chosen as to completely activate the network. See the discussion in Sect. 3.7.2

As argued in Sects. 3.5 and 3.6, the combination of synchronization and acceleration introduces a competition for coherence among the stored patterns. At the end of Sect. 3.6.3, it was concluded that each pattern p participates in the competition with a strength determined by $\lambda_p n_p$. This is confirmed with Fig. 3.5. Only pattern $p = 1$ takes a state of enduring coherence, it is winning the competition because clearly $\lambda_1 n_1 = 15/P > \lambda_q n_q$ for $q = 2, \ldots, P$ ($P = 6$). The other patterns return to states of decoherence, their trajectories are of a cycloidal form (see [5, Sect. 3.1] for an explanation of the denotation *cycloidal*, used in the sense of "cycloid-like"). We refer to a pattern that is winning this competition, like pattern $p = 1$ in Fig. 3.5, as dominating.

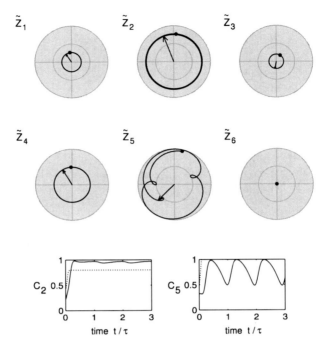

Fig. 3.6 Example 3. The example demonstrates how the activity of the network, that is, the set of on- and off-states affects the competition for coherence. The only difference with Example 1 is a different set of inputs I_k that imply that the network is only partially activated. (The other examples use inputs I_k that result in complete activation of the network, that is, all units become on-state, $V_k \simeq 1$.) To make the partial activations visible, the disks now display $\widetilde{Z}_p = R_p \exp(i\Psi_p)$, where $R_p = A_p C_p$, from $t = 0.5\tau$ to $t = 3\tau$. As described in Sect. 3.7.3, the competition for coherence reduces to patterns $p = 2$ and $p = 5$. The time courses of their pattern coherences (*solid lines*) and activities (*broken lines*) are shown at the *bottom*. In accordance with $\lambda_2 n_2 > \lambda_5 n_5$, pattern $p = 2$ is found to be the dominating pattern; see also the discussion in Sect. 3.7.3

3.7.3 Example 3: Relevance of Classical Activity

In the context of the PCM given by Eqs. 3.11a, 3.11b, 3.12a, 3.12b, and 3.17, classical and temporal coding take complementary roles. This was discussed in detail in [7]. The classical dynamics may determine whether the neural units take on- or off-states, thereby giving values to the quantities n_p for each pattern. The dynamics does then imply a competition for coherence on the remaining on-state units. A pattern p with sufficiently larger $\lambda_p n_p$ may then win the competition, thereby being segregated with respect to pattern coherence. (Notice, however, also the remarks in Sect. 3.7.5.)

Example 3 serves to illustrate this point. Instead of using the default input, we use an input that suppresses the activity of some units so that the units of patterns $q = 1, 3, 4$ become off-state if the unit does not also participate in patterns $p = 2$ or $p = 5$. Moreover, two units of pattern $p = 2$ become off-state (this was arranged

3 Temporal Coding Is Not Only About Cooperation—It Is Also About Competition

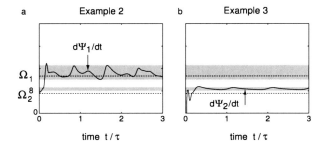

Fig. 3.7 Examples 2 and 3. The figure illustrates the pattern-frequency-correspondence by showing the phase velocities $d\Psi_p/dt$ of the dominating patterns p in comparison to the corresponding pure pattern frequencies (displayed as *dotted horizontal lines*). (**a**) In case of Example 2, the pattern phase velocity $d\Psi_1/dt$ of the dominating pattern $p = 1$ is close to the pure pattern frequency Ω_1 (we abbreviate $\Omega_p^{N_p} = \Omega_p$). (**b**) In case of Example 3, the dominating pattern is $p = 2$. This is partially activated with $n_2 = 8$ on-state units (the complete pattern has $N_2 = 10$ units k with $\xi_k^2 = 1$). The pattern frequency $d\Psi_2/dt$ is in a range that is clearly distinct from that of the dominating pattern in case of Example 2. It is found to be closer to the pure pattern frequency $\Omega_p^{n_p} = \Omega_2^8$. The differences between actual phase velocities and pure pattern frequencies is due to the idealization that is assumed for determining the $\Omega_p^{n_p}$; see Sect. 3.7.3

to illustrate that also partially active patterns may be dominating). Pattern $q = 6$ gets completely off-state. This reduces the competition for coherence to patterns $p = 2$ and $p = 5$. In consequence, we find that pattern $p = 2$ is dominating since $\lambda_2 n_2 = 8/P > \lambda_5 n_5 = 4/P$; see Fig. 3.6.

In Sect. 3.6, the pure-pattern-frequency spectrum was used as a means to understand that a competition for coherence arises. This spectrum was the result of considering an idealized scenario. With Examples 2 and 3, the dominating patterns are $p = 1$ and $p = 2$, respectively. Their dynamics do not represent pure pattern states, since the acceleration due to overlapping patterns is not vanishing. Nevertheless, we find some correspondence between the phase velocities of the dominating patterns and the corresponding idealized pure pattern frequencies; see Fig. 3.7. This should confirm the heuristic arguments given in Sect. 3.6.

3.7.4 Example 4: Relevance of Pattern Weights

As stated in Sect. 3.6.3, the quantities $\lambda_p n_p$ determine the dominating patterns. A comparison of Examples 2 and 3 allowed to demonstrate the dependency on the number of on-state units n_p. Let us now return to the situation of Example 2, where the inputs are such that the network gets completely activated, that is, $n_p = N_p$ for each pattern. We consider Example 4 that differs from Example 2 only by using non-identical values for the pattern weights that enter the Hebbian couplings in Eqs. 3.2 and 3.8. While the default values are $\lambda_p = 1/P$ for the patterns $p = 1, \ldots, P$, we now assume with Example 4 that the weight of pattern $p = 2$ is higher:

$$(\lambda_1, \ldots, \lambda_P) = (1, 2, 1, 1, 1, 1)/(P + 1), \qquad (3.25)$$

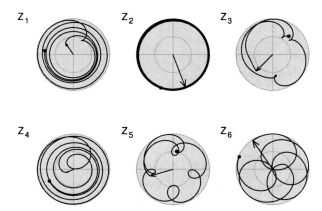

Fig. 3.8 Example 4. This example uses the same parameters, stored patterns and initial values as Example 2, except for the choice given by Eq. 3.25. This gives higher weight λ_p to pattern $p = 2$ so that $\lambda_2 n_2 > \lambda_q n_q$, for $q = 1, 3, 4, 5, 6$. In consequence, pattern $p = 2$ is dominating, it takes the state of enduring coherence, while the other patterns show cycloidal trajectories. The panels show $Z_p = C_p \exp(i\Psi_p)$, for $t = 0.5\tau$ to $t = 3\tau$, as before. The time course of activation and coherence are not displayed (as with Example 2, the network gets completely activated). See the discussion in Sect. 3.7.4

where $P = 6$. In consequence, one obtains $\lambda_2 n_2 > \lambda_q n_q$, for $q = 1, 3, 4, 5, 6$. In particular, $\lambda_2 n_2 = 2n_2/(P+1) = 20/7 > \lambda_1 n_1 = n_1/(P+1) = 15/7$. This explains why the dominating pattern is now $p = 2$; see Fig. 3.8.

3.7.5 Example 5: Assemblies Through Binding of Patterns

The final example should clarify two points that were underemphasized with the foregoing discussions. First, winning assemblies may consist of more than one pattern. Second and related to the first point, including acceleration does not only imply the competition for coherence, it also affects cooperation with respect to the notion of binding, where binding is understood as phase-locking of oscillators. Binding is usually identified with locking of single neural units. In the presence of acceleration, this notion of binding may be extended to locking of coherent patterns [8].

The number of patterns that form the assembly depends on the synchronization strength σ in comparison to the acceleration parameter $\tau\omega_2$. This should be obvious by noticing that a value of σ that is very much larger than $\tau\omega_2$ will lead to global synchronization. It is then interesting to see that increasing the value of σ (or decreasing the value of $\tau\omega_2$) defines a road towards global coherence that reveals a stability hierarchy of the patterns. With Example 5, we demonstrate only one case, where the value of σ is doubled in comparison to the other examples (the other parameters have default values of Example 2). In consequence, we find that the winning assembly consists of three patterns; see Fig. 3.9. See [8] for a more

3 Temporal Coding Is Not Only About Cooperation—It Is Also About Competition

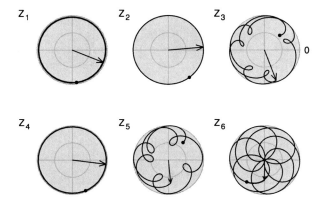

Fig. 3.9 Example 5. Winning assemblies may consist of more than one pattern. Here, the winning assembly consists of patterns $p = 1, 2, 4$. This example uses the same parameters, stored patterns and initial values as Example 2, except for a doubled value of synchronization strength, $\sigma = 2\tau_\theta \omega_2$. This phenomenon allows to extend the notion of binding from phase-locking of single oscillators to phase-locking of patterns. See the discussion in Sect. 3.7.5

extensive discussion of this aspect, including examples that illustrate the mentioned stability hierarchy.

See also the remarks and references in Sect. 3.8 on the possibility of assemblies formed through hierarchical couplings of patterns.

3.8 Outlook

3.8.1 Learning and Inclusion of Inhibitory Couplings

As mentioned above, it is remarkable that the competition for coherence is based on excitatory Hebbian couplings alone, without need for including inhibitory couplings. This points to at least two directions of future research. Firstly, being based on Hebbian couplings, the competition may be subject to learning. Thus, appropriate learning procedures may be implemented. Secondly, the fact that acceleration is based on excitatory couplings leaves freedom to include inhibition in a manner that supports the mentioned complementarity of classical and temporal coding (illustrated with Example 3 in comparison to Example 2).

A first approach to include inhibition was given in [6]. Another approach to implement inhibition, more oriented along the biological example, was given in [12]. This latter approach still needs to be combined with the acceleration mechanism. The general picture one may be aiming at is that inhibitory couplings serve to establish a competition-for-activity that is combined with the competition-for-coherence in a complementary manner [12]; see also the remarks in [8, Sects. 1 and 4].

Recently, a new approach towards including inhibition was proposed that is strongly inspired by properties of brain dynamics, in particular, by the fact that inhibitory oscillations play a central role in the cortical gamma oscillations; see, for example, [34] and references therein for work on the generation of the gamma oscillations in relation to inhibitory oscillations (the relevant kind of oscillations was referred to as pyramidal-interneuron network gamma oscillations (PING) in [35]). The new mechanism, referred to as "windowing mechanism" and described in [36], was also inspired by the properties that arise with the PCMs as discussed in the foregoing chapters.

Briefly speaking, the windowing mechanism assumes, firstly, that the inhibitory rhythm is generated by the rhythms of the excitatory units (through columnar couplings). As we saw in the foregoing sections—see, for example, Fig. 3.7—different parts of the excitatory network may tend to imprint different rhythms on the inhibitory units. Only the patterns that win the competition for coherence may tend to imprint a coherent rhythm as, for example, pattern $p = 2$ in Fig. 3.8. Secondly, these inhibitory units act back on the excitatory units in a rather diffuse manner, constituting an inhibitory pool. The central approach of the windowing mechanism is to use the fact that the part (patterns) of the network that is dominating the competition for coherence may also dominate the (imprinted) inhibitory rhythm so that—in the feedback of the inhibitory pool onto the excitatory units—only the dominating patterns are compatible with the inhibitory rhythm. This may lead to a selective and self-amplifying form of inhibition. It could serve as a readout mechanism for the patterns that win the competition for coherence. See [36] for a first report on such a mechanism (there, we also motivate the denotation as "windowing mechanism"). Work into this direction is in progress.

3.8.2 Hierarchical Processing

Another important research direction is to consider implementing hierarchical, that is, bottom-up and top-down information flow. In analogy to Example 5, one may then attempt to consider assemblies of patterns that are related not through overlap but through association. A proposal how the mechanism presented here may be of relevance in that respect, using a nondiagonal form of the d_{pq} in Eq. 3.2, may be found in [6, Sect. 6]. In this regard, advanced applications may be a goal as guiding principles. For example, with respect to visual object recognition, one may wonder whether the recognition-by-components model, proposed by Irving Biederman [37], could correspond to the framework presented here. The so-called "geons", representing simple geometric primitives, could correspond to the patterns discussed here, while the complete objects would correspond to the assemblies in the sense described above. The combination of several patterns into an assembly could possibly be realized, firstly, in a manner like the one illustrated with Example 5 above and, secondly, through hierarchical processing. In that respect, see also the remarks in [8, Sect. 4].

3.8.3 Mathematical Aspects

The discussed PCM may also be interesting from a purely mathematical perspective. It appears to show a rich bifurcation structure (for example, the stability hierarchy mentioned with Example 5 and discussed more extensively in [8]) that deserves further studies. The arguments for the dynamics given above are merely heuristic and deeper analytical understanding may be helpful also for modeling. Moreover, higher-mode couplings have been described in [4, Sect. 2.4], but their effect still needs to be studied. Notice in this respect, the study of cluster states arising from higher-mode couplings in the context of another phase model given in [38].

3.8.4 Biological Relevance and Time Delays

Finally, the relevance of the proposed mechanisms for describing brain phenomena remains to be clarified. In this regard, the described oscillators may be interpreted as representing groups of neurons, possibly related to a columnar structure [12], and also averaged with respect to time, possible in the sense of pulse-averaging that was described in [39]. It was speculated in [7] that the observed "recoding of excitatory drive into phase-shifts" in relation to the gamma cycle may be related to the acceleration mechanism; see [40] for a review of the gamma cycle as a building block to realize temporal coding in brain dynamics. Related to this context, see also [13] as an approach towards implementing synchronization with close-to-zero time-lag in networks of coupled phase model oscillators despite substantial time delays between the oscillating units.

It is now known that the inhibitory system is crucially involved in generating the gamma rhythm; see the review in [40]. The inhibitory effect therefore combines with the mentioned recoding of excitatory drive into phase-shifts, making it difficult to seperate the different contribution from an experimental perspective. Nevertheless, the recent progress in studies of systematic dependencies between phase-shifts and stimulus properties is remarkable; see [41–43] and references therein. Correspondingly, also from the modeling perspective, the topic of biological interpretation meets the necessity to include inhibition, mentioned at the beginning of this outlook. Given the relevance of the inhibitory units, the study of neurophysiologically observable consequences should probably rather concentrate on biologically more complete networks structures—perhaps like the ones with "windowing mechanism" mentioned in Sect. 3.8.1.

3.9 Summary

Recent progress towards modeling temporal coding was reviewed and an outlook was given on future research directions. The review and outlook concentrated on new perspectives that are based on simple network models of coupled phase model oscillators by complementing synchronization with the so-called acceleration mechanism. We gave simple arguments that described the benefit that temporal coding may hold, pointing to the need of establishing a competition for coherence among the patterns. The patterns were assumed to be stored according to Hebbian memory. It was argued and demonstrated with examples that such a competition may be implemented by including the acceleration mechanism that lets a unit oscillate with higher phase-velocity in case of stronger and/or more coherent input from the connected units. Here, we referred to models using this mechanism as Patterned Coherence Models (PCMs). Temporal assemblies were then identified with the winning configuration, a set of one or more phase-locked patterns. Thereby, acceleration does not only introduce competition for coherence, it also allows for a redefinition of cooperation: binding may not only refer to binding of oscillatory units but also to binding of patterns into temporal assemblies.

Acknowledgements The author is thankful to Christoph von der Malsburg for inspiring and helpful discussions on self-organization mechanisms in general and temporal coding in particular. Moreover, it is a pleasure to thank Danko Nikolić, Martha Havenith and Gordon Pipa from the Max-Plack-Institute for Brain Research, Frankfurt, for insightful discussions on temporal coding with respect to observed brain dynamics.

References

1. Roskies AL (1999) Reviews on the binding problem. Neuron 24:7–9
2. von der Malsburg C (1999) The what and why of binding: the modeler's perspective. Neuron 24:95–104
3. Haken H (1980) Synergetics. Are cooperative phenomena governed by universal principles? Naturwissenschaften 67:121–128
4. Burwick T (2007) Oscillatory neural networks with self-organized segmentation of overlapping patterns. Neural Comput 19:2093–2123
5. Burwick T (2008) Temporal coding: assembly formation through constructive interference. Neural Comput 20:1796–1820
6. Burwick T (2008) Temporal coding with synchronization and acceleration as complementary mechanisms. Neurocomputing 71:1121–1133
7. Burwick T (2008) Temporal coding: the relevance of classical activity, its relation to pattern frequency bands, and a remark on recoding of excitatory drive into phase shifts. Biosystems 94:75–86
8. Burwick T (2008) Assemblies as phase-locked pattern sets that collectively win the competition for coherence. In: Proceedings of 18th international conference on artificial neural networks (ICANN'2008), Part II, Prague, Czech Republic, 3–6 September. Lecture notes in computer science. Springer, Berlin, pp 617–626
9. Burwick T (2006) Oscillatory networks: pattern recognition without a superposition problem. Neural Comput 18:356–380
10. Cohen MA, Grossberg S (1983) Absolute stability of global pattern formation and parallel memory storage by competitive neural networks. IEEE Trans Syst Man Cybern SMC-13:815–826
11. Hopfield JJ (1984) Neurons with graded response have collective computational properties like those of two-state neurons. Proc Natl Acad Sci USA 81:3088–3092

12. Burwick T (2009) On the relevance of local synchronization for establishing a winner-take-all functionality of the gamma cycle. Neurocomputing 72:1525–1533
13. Burwick T (2009) Zero phase-lag synchronization through short-term modulations. In: Proceedings of the 17th European symposium on artificial neural networks (ESANN'2009), Bruges, Belgium, 22–24 April, pp 403–408
14. Haykin S (1999) Neural networks—a comprehensive foundation, 2nd edn. Prentice Hall, New York
15. Sakagushi H, Shinomoto S, Kuramoto Y (1988) Mutual entrainment in oscillator lattices with nonvariational type interaction. Prog Theor Phys 79:1069–1079
16. von der Malsburg C (1981) The correlation theory of brain function. Max-Planck Institute for Biophysical Chemistry, Internal Report 81-2
17. von der Malsburg C (1983) How are nervous structures organized. In: Basar E, Flor H, Haken H, Mandell AJ (eds) Synergetics of the brain. Springer series in synergentics, vol 23. Springer, Berlin, pp 238–249
18. von der Malsburg C (1986) Am I thinking assemblies. In: Palm G, Aertsen A (eds) Brain theory, October 1–4, 1984. Proceedings of first Trieste meeting on brain theory. Springer, Berlin, pp 161–176
19. Shimizu H, Yamaguchi Y, Tsuda I, Yano M (1985) Pattern recognition based on holonic information dynamics: towards synergetic computers. In: Haken H (ed) Complex systems—operational approaches in neurobiology, physics, and computers. Springer series in synergetics. Springer, Berlin, pp 225–239
20. von der Malsburg C, Schneider W (1986) A neural cocktail-party processor. Biol Cybern 54:29–40
21. Baird B (1986) Nonlinear dynamics of pattern formation and pattern recognition in the rabbit olfactory bulb. Physica D 22:242–252
22. Freeman WJ, Yao Y, Burke B (1988) Central pattern generating and recognizing in olfactory bulb: a correlation learning rule. Neural Netw 1:277–288
23. Li Z, Hopfield JJ (1989) Modeling the olfactory bulb and its neural oscillatory processings. Biol Cybern 61:379–392
24. Kuramoto Y (1984) Chemical oscillations, waves, and turbulence. Springer series in synergetics. Springer, Berlin
25. Acebrón JA, Bonilla LL, Vicente CJP, Ritort F, Spigler R (2005) The Kuramoto model: a simple paradigma for synchronization phenomena. Rev Mod Phys 77:137–185
26. Hoppensteadt FC, Izhikevich EM (1997) Weakly connected neural networks. Springer, New York
27. Hirose A (ed) (2003) Complex-valued neural networks. World Scientific, Singapore
28. Hirose A (2006) Complex-valued neural networks. Springer, Berlin
29. Rosenblatt F (1961) Principles of neurodynamics: perceptrons and the theory of brain mechanism. Spartan Books, Washington
30. Winfree AT (1980) The geometry of biological time. Springer, New York
31. Ermentrout G, Kopell N (1984) Frequency plateaus in a chain of weakly coupled oscillators. SIAM J Appl Math 15:215–237
32. Borisyuk RM, Kazanovich YB (1994) Synchronization in a neural network of phase oscillators with the central element. Biol Cybern 71:177–185
33. Borisyuk RM, Kazanovich YB (2004) Oscillatory model of attention-guided object selection and novelty detection. Neural Netw 17:899–915
34. Börgers C, Kopell NJ (2005) Effects of noisy drive on rhythms in networks of excitatory and inhibitory neurons. Neural Comput 17:557–608
35. Whittington MA, Traub RD, Kopell N, Ermentrout B, Buhl EH (2000) Inhibition-based rhythms: experimental and mathematical observations on network dynamics. Int J Psychophysiol 38:315–336
36. Burwick T (2011) Pattern recognition through compatibility of excitatory and inhibitory rhythms. Neurocomputing 74:1315–1328
37. Biederman I (1987) Recognition-by-components: a theory of human image understanding. Psychol Rev 94:115–147

38. Tass PA (1999) Phase resetting in medicine and biology. Springer, Berlin
39. Haken H (2002) Brain dynamics: synchronization and activity patterns in pulse-coupled neural nets with delays and noise. Springer, Berlin
40. Fries P, Nicolić D, Singer W (2007) The gamma cycle. Trends Neurosci 30(7):309–316
41. Schneider G, Havenith M, Nikolić D (2006) Spatio-temporal structure in large neuronal networks detected from cross correlation. Neural Comput 18:2387–2413
42. Nikolić D (2007) Non-parametric detection of temporal order across pairwise measurements of time delays. J Comput Neurosci 22:5–19
43. Havenith M, Zemmar A, Yu S, Baudrexel S, Singer W, Nikolić D (2009) Measuring sub-millisecond delays in spiking activity with millisecond time-bins. Neurosci Lett 450:296–300

Chapter 4
Using Non-oscillatory Dynamics to Disambiguate Pattern Mixtures

Tsvi Achler

Abstract This chapter describes a model which takes advantage of the time domain through feedback iterations that improve recognition and help disambiguate pattern mixtures. This model belongs to a class of models called generative models.

It is based on the hypothesis that recognition centers of the brain reconstruct an internal copy of inputs using knowledge the brain has previously accumulated. Subsequently, it minimizes the error between the internal copy and the input from the environment.

This model is distinguished from other generative models which are unsupervised and represent early visual processing. This approach utilizes generative reconstruction in a supervised paradigm, implementing later sensory recognition processing.

This chapter shows how this paradigm takes advantage of the time domain, provides a convenient way to store recognition information, and avoids some of the difficulties associated with traditional feedforward classification models.

4.1 Introduction

To survive animals must interpret mixtures of patterns (scenes) quickly, for example, find the best escape path from an encircling predator pack; identify food in a cluttered forest. Scenes are commonly formed from combinations of previously learned patterns. Pattern mixtures are found in AI scenarios such as scene understanding, multiple voice recognition, and robot disambiguation of sensory information.

Difficulty in processing mixtures of patterns is a fundamental problem in connectionist networks. Difficulties are described even in the earliest connectionist literature as: "the binding problem" and "superposition catastrophe". Connectionist networks do not generalize well with pattern mixtures. As a consequence, the number of required nodes or training epochs grows exponentially as the number of patterns increases, causing a combinatorial explosion [6, 30, 32].

T. Achler (✉)
Siebel Center, University of Illinois Urbana-Champaign, 201 N. Goodwin Ave, Urbana, IL 61801, USA
e-mail: achler@illinois.edu

To quantify the implications, let n represent the number of possible output nodes or representations. Let k represent the number of mixed patterns. The number of possible mixed pattern combinations is given by:

$$\binom{n}{k} = \frac{n!}{k!(n-k)!}. \tag{4.1}$$

If a network contains $n = 5{,}000$ patterns, there are about *12 million* two pattern combinations $k = 2$, and *21 billion* three pattern $k = 3$ combinations possible. If a network contains $n = 100{,}000$ patterns, there are *5 billion* $k = 2$ and *167 trillion* $k = 3$ combinations and so on. The brain must process mixed patterns more efficiently because it cannot train for each possible combination for the simple reason that there are not enough resources, such as neurons, let alone the time needed for such preparation.

Another way to quantify this difficulty is to compare the training and test distributions. Weight parameters are determined through the training data. Determination of weights requires the training distribution to be similar to the testing distribution. Similar distributions allow the correlation between input features and outcomes to be determined through a training set, and learning to occur. However, the training distribution is commonly violated in the natural (test) environment, such as a scene or overlapping patterns. If a network is trained on patterns A and B presented by themselves and they appear simultaneously, this is outside the training distribution. As previously discussed, training for every pair of possible patterns (or triplets, quadruplets, etc.) is combinatorially impractical.

It is often assumed that most combinatorial limitations can be overcome by isolating or segmenting patterns individually for recognition by utilizing their spatial location [11, 19–21, 24, 27, 28, 34, 38]. However, segmenting patterns in cluttered scenarios is difficult because prior recognition is often required for correct localization. Moreover, some modalities such as olfaction or taste do not allow for good spatial localization, because the sensory information is broadly scattered in space [10]. Yet emulating recognition performance in olfaction remains as difficult if not more difficult than spatial-rich modalities such as vision [8, 29].

Methods incorporating oscillatory dynamics, for example [31, 36] and this book, have been successful in differentiating components of patterns within a scene and partially addressing these problems. However, most methods require some a-priori segmentation.

This work does not make any assumptions about space, and patterns are not isolated or segmented in space. A dynamic model is proposed to address binding and pattern mixture processing issues that is independent of oscillations, synchronization, or spike timing. Yet it uses temporal dynamics. This Regulatory Feedback model focuses on the contribution of top-down inhibitory structures. It works by cycling activation between inputs and outputs. Inputs activate contending representations which in turn inhibit their own representative inputs. In general, inputs that are utilized by multiple representations are more ambiguous (not as unique) compared to those utilized by single representations. Ambiguous inputs are inhibited by multiple outputs and become less relevant. The inhibited inputs then affect representation activity, which again affects inputs. The cycling is repeated until a steady state is reached. This method facilitates simultaneous evaluation of representations and determines sets of representations that best fit the whole "scene". The implementation of non-oscillatory feedback dynamics for separating patterns is described in detail and key examples are demonstrated by simulations.

4.1.1 Evidence for Top-Down Feedback Regulation in the Brain

Regulation of an input by the outputs it activates is referred to as regulatory feedback. Sensory processing regions of the brain (e.g. thalamus, olfactory bulb, sensory cortex) have a massive amount of reentrant top-down feedback pathways which regulate bottom-up processing. Furthermore, a large variety of brain mechanisms can be found where inputs are regulated, even at the level of synapses [9, 15, 18].

Anatomical studies provide evidence for top-down regulatory connections. For example, in the thalamo-cortical axis, thalamic relay neurons (inputs) innervate cortical pyramidal cells (outputs). These cortical neurons feed-back and innervate the thalamic reticular nucleus. The neurons from the thalamic reticular nucleus inhibit the relay neurons. Feedback must be very tightly controlled because there are more feed-back connections than feed-forward connections [13, 14, 16, 22, 23].

Homeostatic plasticity provides another form of evidence. In homeostatic plasticity, pre-synaptic cells (inputs) will change their activation properties to restore homeostasis (fixed input to output activity levels), between pre- and post-synaptic cells (between inputs and outputs). This requires top-down (output to input) regulatory feedback targeting pre-synaptic neurons [25, 35]. Although the timescale is slower, homeostatic plasticity provides evidence that connections required for regulatory feedback exist. The real time contribution of such connections can facilitate conservation of information.

4.1.2 Difficulties with Large Scale Top-Down Regulatory Feedback

Unfortunately, large-scale amounts of regulatory feedback are difficult to incorporate into models, because the mathematics become highly nonlinear and difficult to analyze. Control theory, a field in engineering, is dedicated to the study of feedback properties. Within this field, regulatory feedback would be considered to as a type of negative feedback. However, during recognition, most neural models focus on more tractable configurations, such as parameterized feed-forward connections and lateral inhibition (see Fig. 4.1).

4.2 Regulatory Feedback Networks

The feedback regulation algorithm is realized as a two layer network with fuzzy-type input features x, output nodes y, and no hidden units. Regulatory feedback is demonstrated here using nodes with binary connections to better appreciate the contribution of dynamics. The point of implementing binary connections is to appre-

Fig. 4.1 Comparison of mechanisms utilized in recognition. Feedforward connections: neural networks and support vector machines. Lateral inhibition: winner-take-all. Feedback regulation: top-down self-inhibition, negative feedback

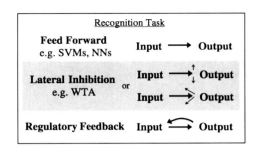

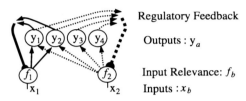

Regulatory Feedback

Outputs: y_a

Input Relevance: f_b

Inputs: x_b

Fig. 4.2 Example configuration of symmetric self-regulation. If x_1 activates $y_1 \& y_2$, then feedback from $y_1 \& y_2$ regulates x_1. Similarly, if x_2 activates y_1, y_2, $y_3 \& y_4$, then feedback from y_1, y_2, $y_3 \& y_4$ regulates x_2. Thus, for this example: $M_1 = \{y_1, y_2\}$, $M_2 = \{y_1, y_2, y_3, y_4\}$, $G_1 = G_2 = \{x_1, x_2\}$, $G_3 = G_4 = \{x_2\}$, $g_3 = g_4 = 1$, $g_1 = g_2 = 2$

ciate limits of feedforward networks with analagous binary connections. Dynamic recognition is possible even if all input features are connected equally to associated output nodes (in a binary fashion). In regulatory feedback networks input relevance f is determined at the time of recognition; for example, value $f = 1$ is salient, but 2 is twice as salient and 0.5 is half.

Each output node y_a is defined by set of feedforward binary connections G_a from input x's. It also has a set of symmetrical feedback connections M_a that implement negative feedback, modulating relevance (Fig. 4.2). Connectivity strengths are uniform here, which means all x's that project to a y have the same weight. Similarly all of a y's feedback connections to its x's have the same weight. Let's label M_b the connections that return to an input x_b. Let's label f_b the relevance value of input x_b. Then the activity of the output node is determined by:

$$y_a(t+dt) = \frac{y_a(t) \cdot dt}{g_a} \sum_{i \in G_a} f_i \qquad (4.2)$$

with regulatory feedback affecting every input b

$$f_b = \frac{x_b}{Y_b} \qquad (4.3)$$

through feedback term.

$$Y_b = \sum_{j \in M_b} y_j(t). \qquad (4.4)$$

For an output node y_a, set G_a denotes all its input connections and g_a denotes the number of these connections. For an input x_b, set M_b denotes all output nodes connected to it. The feedback from output nodes to input x_b is Y_b. The value Y_b represents the sum of all of the output nodes that use x_b. The input relevance is f_b, determined by feedback composed of the sum of all the outputs using that input.

Feedback modifies relevance values. The modified relevance values are redistributed to the network, determining new output activations. This modification process is repeated iteratively to dynamically determine final relevance of inputs and recognition. The iterative nature allows inference during the recognition phase.

The networks dynamically test recognition of representations by (1) evaluating outputs based on inputs and their relevance, (2) modifying the next state of input relevance based on output use of the inputs, (3) reevaluating outputs based on the new input relevance. Steps 1–3 represent a cycle which can be iteratively repeated.

Modularity and Scalability This structure allows modularity and scalability because each node only connects to its own inputs. Thus, output nodes do not directly connect with the other output nodes. Instead they interact indirectly through shared inputs. As a result, an addition of a new node to the network only requires that it forms symmetrical connections to its inputs (an input-to-output connection paired with an equivalent output-to-input connection). Subsequently, the number of connections of a node is independent of the size or composition of the network. Yet the networks make complex recognition decisions based on distributed processing, as demonstrated in this work.

4.2.1 Description of Function

The networks dynamically balance the ratio of input and output of activity. Suppose an input provides information x_a to its outputs Y_a. These outputs feed back in unison to the input x_a and regulate its relevance in a 'shunting' fashion $f_a = x_a/Y_a$, whereby inhibition does not completely eliminate an input's activity but reduces it. The tight association creates a situation where the information x_a can be fully expressed to the output layer only if $Y_a = 1$ (which occurs when inputs and outputs are matched). If several outputs are overly active, no output will receive the full activity of x_a because $Y_a > 1$ thus $f_a < x_a$. Conversely if x_a is not appropriately represented by the network then $Y_a < 1$ and the input is boosted $f_a > x_a$. In that case outputs will recieve more input activity than designated by x_a. This regulation occurs for every input–output interaction.

Additionally, each output node strives for a total activation of 1. Thus, if an output cell has N inputs each connection contributes $1/N$ activity. This normalizes output activity and promotes the relevance value of 1 as a steady state fixed point.

4.2.2 General Limits, Stability, Steady State, & Simulations

The outputs y are shown analytically to be bounded between zero and x values [1, 4] and this class of equations settle to a non-oscillatory steady state equilibrium [26].

In this network, all variables are limited to positive values. Thus, the values of y cannot become negative and have a lower bound of 0. The upper values of y are bounded by the maximal input given value $x_{\max}$. y_a activity will be greatest if its relevance f_{G_a} is maximized. Relevance will be maximized if nodes that share its inputs are not active $M_a \notin y_a = 0$. Assuming this is the case then the equation simplifies to:

$$y_a(t+dt) = \frac{y_a(t)}{g_a} \sum_{i \in G_a} f_i$$

$$\leq \frac{y_a(t)}{g_a} \sum_{i \in G_a} \left(\frac{x_{\max}}{y_i(t)+0} \right)$$

$$= \frac{1}{g_a} \sum_{i \in G_a} (x_{\max}) = \frac{x_{\max} \cdot g_a}{g_a} = x_{\max}.$$

If $x_{\max}$ is bounded by 1, then y_a is bounded by 1. The values are bounded by positive numbers between zero and $x_{\max}$, satisfying boundary conditions [6]. Furthermore as $dt \to 0$, a Lyapunov function can be written. This indicates that the networks will settle to a steady state and not display chaotic oscillations. Numerical simulations also show the equations are well behaved.

In numerical simulations, y_a is constrained to a minimum of epsilon ε to avoid a divide-by-zero error (ε is a very small real number; 10^{-7} in practice). Steady state is defined when all output node values change by less than 0.0001 in a simulation cycle.

4.2.3 Simple Two Node Demonstration

A simple toy two node configuration (Fig. 4.3, Example 1) is presented to illustrate the workings and dynamics of the model. Two nodes are connected such that the first node y_1 has one input (x_1) which overlaps with the second node y_2. y_2 has two inputs ($x_1 \& x_2$). Since y_1 has one input, $g_1 = 1$ in the equation for y_1. Since y_2 has two inputs $g_2 = 2$ in the equation for y_2 and each input to y_2 contributes one-half. x_1 projects to both $y_1 \& y_2$, thus f_1 receives feedback from both $y_1 \& y_2$. x_2 projects only to y_2 so f_2 receives feedback only from y_2.

Such compositions can be created in a modular fashion. Suppose one node represents the pattern associated with the letter P and another node represents the pattern associated with the letter R. R shares some inputs with P. These nodes can be combined into one network, by just connecting the network and ignoring this overlap. This defines a functioning network without formally learning how nodes R and P should interact with each other. Whether the features in common with R and P should predominantly support R or P or both is determined dynamically during testing (via activation and feedback), not training (i.e., though an a-priori predetermined weight).

4 Using Non-oscillatory Dynamics to Disambiguate Pattern Mixtures

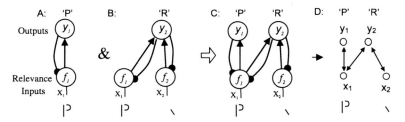

Fig. 4.3 (A–C), Example 1: Modular nodes y_1 and y_2 (A & B, respectively) can be simply combined to form a coordinated network (C). No modifications to the base nodes or new connections are needed. The nodes interact with each other at the common input, x_1. Since the feedforward and feedback connections are symmetric, the network can be drawn using bidirectional connections (D)

Equations of the above configuration are given by:

$$y_1(t+dt) = \frac{y_1(t)x_1}{y_1(t)+y_2(t)}, \tag{4.5}$$

$$y_2(t+dt) = \frac{y_2(t)}{2}\left(\frac{x_1}{y_1(t)+y_2(t)} + \frac{x_2}{y_2(t)}\right). \tag{4.6}$$

Stepping through iterations yields an intuition about the algorithm, see Fig. 4.4. Lets assume $x_1 = x_2 = 1$, with the initial activity of y_1 and y_2 assumed to be 0.01 ($y_1 = 0.01$ and $y_2 = 0.01$ at $t < 1$). $t = 0$ represents the initial condition. At $t = 1$ *backward* the activity of y_1 & y_2 are projected back to the inputs. Both s_1 and s_2 are boosted because representations that use x_1 & x_2 are not very active (initial values are 0.01). Thus, the input relevance (f_1 and f_2) of x_1 & x_2 are boosted. Note that f_2 is boosted more than f_1 because two nodes use input f_1. At $t = 1$ *forward* the inputs modulated by salience are projected to the output nodes. Note that both y_1 and y_2 gain activation. The new activation of the output node is a function of the node's previous activity and the activity of the inputs normalized by the number of node processes (g_a). At $t = 2$ *backward* the activity of y_1 & y_2 are projected back to the inputs again. This time the output nodes are more active so f_1 & f_2 values are smaller. From $t = 2$ *forward* to $t \to \infty$ this trend continues, reducing y_1 activity while increasing y_2 activity. At $t = \infty$, the steady state values of y_1 becomes 0 and y_2 becomes 1.

The solutions at steady state can be derived for this simple example by setting $y_1(t+dt) = y_1(t)$ and $y_2(t+dt) = y_2(t)$ and solving the equations. The solutions are presented as *(input values)* $\to$ *(output vectors)* where $(x_1, x_2) \to (y_1, y_2)$. The solutions are $(x_1, x_2) \to (x_1 - x_2, x_2)$. The network cannot have negative values (see section on limits and stability) thus if $x_1 \leq x_2$ then $y_1 = 0$ and the equation for y_2 becomes: $y_2 = \frac{x_1+x_2}{2}$. The results for $x_1 = x_2 = 1$ is $(1, 1) \to (0, 1)$. If $x_1 = 1$ and $x_2 = 0$, then y_1 is matched: $(1, 0) \to (1, 0)$.

Thus, if inputs 'P' & '\' are active y_2 wins. This occurs because when both inputs are active, y_1 must compete for all of its inputs with y_2, however y_2 only needs to compete for half of its inputs (the input shared with y_1) and it gets the other half 'free'. This allows y_2 to build up more activity and in doing so inhibit y_1.

Thus, smaller representation completely encompassed by a larger representation becomes inhibited when the inputs of the larger one are present. The smaller repre-

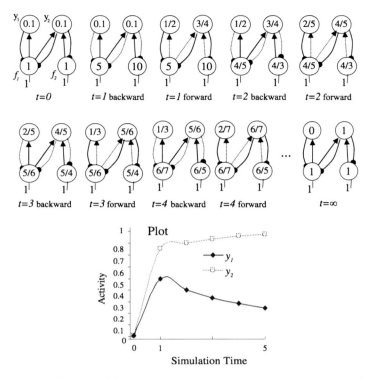

Fig. 4.4 Example 1: Numerical simulation. *Top*: steps in simulation and intermediate values. $t = 0$ initial condition; $t = 1$ feedback contribution at time 1; $t = 1$ feed forward contribution at time 1; $t = 2$ backward next cycle feedback; $t = \infty$ steady state. *Bottom Right*: y_1 & y_2 value plot for first five cycles

sentation is unlikely given features specific only to the large representation. A negative association was implied ('y_1' is unlikely given feature 'x_2') even though it was not directly encoded by training. However, in feedforward networks, such associations must be explicitly trained as demonstrated in Example 2, Fig. 4.8. Such training requires the possible combinations to be present in the training set, resulting in a combinatorial explosion in training. The Binding and Superposition examples demonstrate these training difficulties.

Lastly, though integer inputs are presented, the solutions are defined for positive real x input values and the system is sensitive to numerosity [7]. With twice as much input activity there is twice as much output activity: $(2, 2) \rightarrow (0, 2)$. Patterns are also recognized as mixtures: $(2, 1) \rightarrow (1, 1)$.

4.3 The Binding Problem Demonstration

The binding problem occurs when image features can be interpreted through more than one representation [31, 36]. An intuitive way to describe this problem is

4 Using Non-oscillatory Dynamics to Disambiguate Pattern Mixtures 65

Fig. 4.5 Illusion representing a linearly dependent scenario. In this illusion, multiple labels exist for the same pattern. Every feature can either be interpreted as part of an old women or young woman

through certain visual illusions. In the old woman/young woman illusion of Fig. 4.5, the young woman's cheek is the old woman's nose. A local feature can support different representations based on the overall interpretation of the picture. Though the features are exactly the same, the interpretation is different. In humans, this figure forms an illusion because all features in the image can fit into two representations. Classifiers have similar difficulties but with simpler patterns. If a pattern can be part of two representations, then the networks must determine to which it belongs. Training is used to find optimal interconnection weights for each possible scenario. However, this is not trivial: combinations of patterns and training can grow exponentially.

A simple scenario is presented to show that the regulatory feedback equations behave in a manner which is beneficial towards binding. This scenario demonstrates that compositions of partially overlapping representations can cooperate and/or compete to best represent the input patterns. The equations allow two representations to inhibit a third. The performance of several networks are compared to better illustrate the binding problem within this scenario.

In this composition, three representations are evaluated simultaneously. Example 2 Fig. 4.6 is expanded from Example 1 Fig. 4.3, with a modular addition of a new output node y_3. As in Example 1, y_1 competes for its single input with y_2. However, now y_2 competes for its other input with y_3 and y_3 competes for only one of its inputs.

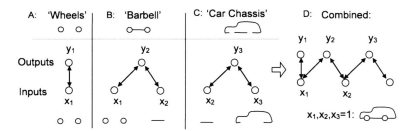

Fig. 4.6 (**A–D**): Modular combination of nodes display binding. Nodes y_1, y_2, y_3 (**A**, **B** & **C**) can be simply combined to form a combined network (**D**). Yet these patterns interact in cooperative and competitive manner finding the most efficient configuration with the least amount of overlap

Fig. 4.7 Simulation of Example 2. Graph of y_1, y_2, & y_3 dynamics given car $(1, 1, 1)$

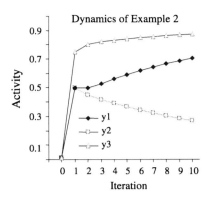

Equation for y_1 remains the same as Example 1. Equations for y_2 and y_3 become:

$$y_2(t+dt) = \frac{y_2(t)}{2}\left(\frac{x_1}{y_1(t)+y_2(t)} + \frac{x_2}{y_2(t)+y_3(t)}\right), \quad (4.7)$$

$$y_3(t+dt) = \frac{y_3(t)}{2}\left(\frac{x_2}{y_2(t)+y_3(t)} + \frac{x_3}{y_3(t)}\right). \quad (4.8)$$

Solving for steady state by setting $y_1(t+dt) = y_1(t)$, $y_2(t+dt) = y_2(t)$, and $y_3(t+dt) = y_3(t)$, the solutions are $y_1 = x_1 - x_2 + x_3$, $y_2 = x_2 - x_3$, $y_3 = x_3$. Thus, $(x_1, x_2, x_3) \rightarrow (y_1 = x_1 - x_2 + x_3, y_2 = x_2 - x_3, y_3 = x_3)$. If $x_3 = 0$ the solution becomes that of for example, 1: $y_1 = x_1 - x_2$ and $y_2 = x_2$. If $x_2 \leq x_3$, then $y_2 = 0$ and the equations become $y_1 = x_1$ and $y_3 = \frac{x_2+x_3}{2}$.

Solutions to particular input activations are:

$(1, 0, 0) \rightarrow (1, 0, 0);$ $\quad (1, 1, 0) \rightarrow (0, 1, 0);$ $\quad (1, 1, 1) \rightarrow (1, 0, 1).$

If only input x_1 is active, y_1 wins. If only inputs x_1 and x_2 are active y_2 wins for the same reasons this occurs in Example 1. However, if inputs x_1, x_2 and x_3 are active then y_1 and y_3 win. The network as a whole chooses the cell or cells that best represent the input pattern with the least amount of competitive overlap.

The dynamics of this network are plotted in Fig. 4.7. In this configuration, y_2 must compete for all of its inputs: x_1 with y_1, x_2 with y_3. y_3 only competes for half of its inputs (input x_2) getting input x_3 'free'. Since y_2 is not getting its other input x_1 'free' it is at a competitive disadvantage to y_3. Together y_1 and y_3, mutually benefit from each other and force y_2 out of competition. Competitive information travels indirectly 'through' the representations. Given active inputs x_1 and $x_2 = 1$, the activity state of y_1 is determined by input x_3 through y_3. If input x_3 is 0, then y_1 becomes inactive. If input x_3 is 1, y_1 becomes active. However, y_3 does not share input x_1 with y_1.

Suppose the inputs represent spatially invariant features where feature x_1 represents circles, x_3 represents the body shape and feature x_2 represents a horizontal bar. y_1 is assigned to represent wheels and thus when it is active, feature x_1 is interpreted as wheels. y_2 represents a barbell O-O composed of a bar adjacent to two round weights (features x_1 and x_2). Note: even though y_2 includes circles (feature

Fig. 4.8 Binding Problem and Classifier Responses given car (1, 1, 1). Of recognition algorithms applied, only regulatory feedback determines solution Y_1 & Y_3 which covers all inputs with the least amount of overlap between representations. Other methods must be further trained to account for the specific binding scenarios

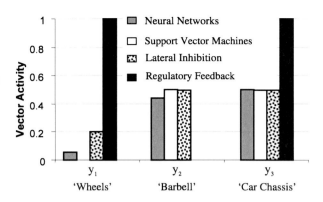

x_1), they do not represent wheels (y_1), they represent barbell weights. Thus, if y_2 is active feature x_1 is interpreted as part of the barbell. y_3 represents a car body without wheels (features x_2 and x_3), where feature x_2 is interpreted as part of the chassis. Now given an image of a car with all features simultaneously (x_1, x_2 and x_3), choosing the barbell (y_2) even though technically a correct representation, is equivalent to a binding error within the wrong context in light of all of the inputs. Thus choosing y_2 given (1, 1, 1) 🚗 is equivalent to choosing the irrelevant features for binding. Most classifiers if not trained otherwise are as likely to choose barbell or car chassis (see Fig. 4.8). In that case, the complete picture is not analyzed in terms of the best fit given all of the information present. Similar to case 1, the most encompassing representations mutually predominate without any special mechanism to adjust the weighting scheme. Thus, the networks are able to evaluate and bind representations in a sensible manner for these triple cell combinations.

Similar to the illusion in Fig. 4.5, multiple representations cooperate to interpret the picture. In the illusion, all of the features either are interpreted as a component of either representation (but not confused as a feature of any other representations).

NN and SVM algorithms are tested using the publicly available *Waikato Environment for Knowledge Analysis* (WEKA) package [37]. The purpose of the WEKA package is to facilitate algorithm comparison and maintains up to date algorithms (2007). The *lateral inhibition* and *feedback regulation* algorithms utilize identical feed-forward connections, but in *lateral inhibition* output nodes compete in a winner-take-all fashion and their fit ranked. Lateral Inhibition combined with ranking of fit (Sect. 4.4) is similar to the Adaptive Resonance Theory (ART) [12].

4.4 Superposition Catastrophe Demonstration

Superposition Catastrophe can be demonstrated in scene-like scenarios, where patterns are presented together without a good way to separate them. The presented toy example demonstrates this problem and how regulatory feedback overcomes it. Like the binding scenario, algorithms are all given the same supervised training set and

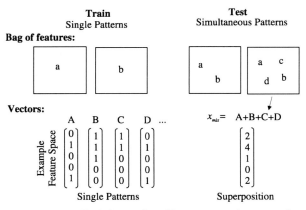

Fig. 4.9 Training with single patterns, testing with pattern mixtures. Test patterns are a superposition mixture of multiple trained patterns. *Top*: Emulation demonstrating mixtures of patterns presented in a bag-of-features feature extractor. *Bottom*: Representative vectors of the patterns. Only the first 5 inputs features of the feature space are shown

test cases. In this case, the training set is composed of randomly generated vector patterns with binary connections. Each input feature has a 50% probability of being associated to each node. There are 512 possible input features. Let n represent the number of possible output nodes. In the examples presented, $n = 26$. These numbers are arbitrary but only 26 random outputs are defined so the output nodes and associated vectors can be labeled $\mathbf{A} \rightarrow \mathbf{Z}$. Networks are trained on single vectors ($\mathbf{A}..\mathbf{Z}$) but tested on simultaneous vectors.

To recognize the 26 patterns, one output node (y) is designated for each pattern. Network "training" is rudimentary. During learning, single patterns are presented to the network, this activates a set of feature inputs. The output node is simply connected to every feature input that is active. Thus, each respective output node is designated to a pattern and is connected to all input features associated to that pattern.

A superposition of pattern mixtures is achieved by adding features of several vectors together generating an intermixed input vector, $\mathbf{x}_{mix}$ (see Fig. 4.9 bottom). Let $\mathbf{x}_{mix}$ represent inputs of pattern mixtures superimposed. For example, $\mathbf{x}_{mix} = \mathbf{A} + \mathbf{B}$ is composed of the sum vectors $\mathbf{A}$ and $\mathbf{B}$. If vectors $\mathbf{A}$ and $\mathbf{B}$ share the same feature, that feature's amplitude in $\mathbf{x}_{mix}$ is the sum of both. This is analogous to a scene composed of multiple objects or an odorant mixture.

One way to implement this is through a bag-of-features type feature extractor or convolution applied to all stimuli in the visual field. Such extractors are similar in spirit to feature extraction found in the primary visual cortex V1, and commonly implemented in cognitive models [33]. $\mathbf{x}_{mix}$ would be the result of placing multiple patterns (far apart) within the bag-of-features extractor (see Fig. 4.9, top).

Networks that are only trained with single patterns are evaluated on scenes composed of two and four pattern mixtures. There are 325 possible tests (e.g., $\mathbf{x}_{mix} = \mathbf{A} + \mathbf{B}, \mathbf{A} + \mathbf{C}, \mathbf{A} + \mathbf{D}\ldots$) with two pattern mixtures, $k = 2$, $n = 26$, see Eq. 4.1. There are 14,950 possible tests (i.e., $\mathbf{x}_{mix} = \mathbf{A} + \mathbf{B} + \mathbf{C} + \mathbf{D}, \mathbf{A} + \mathbf{B} + \mathbf{D} + \mathbf{G}$,

$\mathbf{A} + \mathbf{B} + \mathbf{E} + \mathbf{Q}, \ldots$) with four simultaneous patterns, $k = 4$. All combinations are presented to the network.

At the end the test for a specific combination, the highest ky-values are chosen and their identity is compared to the patterns in $\mathbf{x}_{\text{mix}}$. If the top nodes match the patterns that were used to compose the $\mathbf{x}_{\text{mix}}$, then the appropriate histogram bin (#*number of patterns_matched/k*) is incremented. Suppose $k = 4$, and for a specific test $\mathbf{x}_{\text{mix}} = \mathbf{A} + \mathbf{B} + \mathbf{C} + \mathbf{D}$. If the 4 top active nodes are A, B, C, G, then only three of the top four nodes match the composition in $\mathbf{x}_{\text{mix}}$. The count of bin 3/4 will be incremented.

4.4.1 Results

Neural Networks, Support Vector Machines, and lateral inhibition are again tested. Each method is given single prototypical representations (features of single patterns) for the training phase and tested on multiple representations (features of multiple patterns summed and presented to the network). Only Regulatory Feedback Networks correctly recognize the mixtures (tested up to eight pattern mixtures and score 100%). This occurs despite nonlinear properties of regulatory feedback networks. Regulatory feedback does not require the training and test distributions to be similar [6].

If $k = 4$ (Fig. 4.10 bottom), there are 14,900 possible combinations 4 pattern mixtures. In order for NN, SVMs, and lateral inhibition to match accuracy of feedback inhibition training on the $\sim 15k$ combinations may be required. This is impractical because such numbers grow exponentially with the number of patterns as indicated by Eq. 4.1.

These tests were repeated with various randomly generated patterns formed with various n's (up to 100) and k's (up to 8). As long as (1) no two representations are the same or (2) two representations do not compose a third, regulatory feedback recognized all combinations with 100% success.

4.5 Discussion

These results suggest some of the difficulties within the Superposition Catastrophe and The Binding Problems may originate from the static nature of feedforward models. Dynamic networks allow more robust processing of scenes. This in turn reduces the reliance on scene segmentation while providing more information in order to perform better scene segmentation.

Though combinatorial problems have been associated with feedforward models [30, 32], the feedforward model remains popular because of its flexibility with large parameter spaces.

Fig. 4.10 Superposition catastrophe demonstration. Each network is trained on the same single patterns and tested on the same combinations of the patterns. All possible pattern combinations are tested. The top k active nodes of each network are selected and compared to the presence of original stimuli. The appropriate number_correct/k histogram bin is incremented. *Top*: Results for two pattern mixtures. *Bottom*: Results for four pattern mixtures

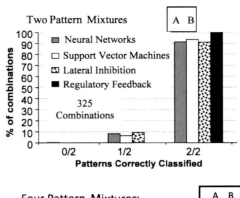

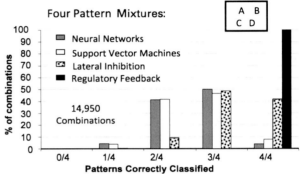

4.5.1 Plasticity Due to Dynamics in Regulatory Feedback Networks

In this demonstration, the patterns **A**..**Z** are linearly dependent. This means that they cannot be expressed as combinations of each other. However if linear dependence does occur regulatory feedback networks still function in a rational manner. For example, if $\mathbf{A} = \mathbf{B}$ (see Fig. 4.11 Case I) then y_A and y_B have same inputs (two labels sharing the same inputs). If that occurs, the network matches the base pattern inputs (vector **A**) and distributes the activation to the dependent patterns. Thus, at steady state $y_A + y_B$ will equal to activation garnered by the **A** configuration. Whether $y_A > y_B$, $y_A < y_B$, or $y_A = y_B$ will be determined by initial values of y_A and y_B. This is a semi-plastic regime that maintains memory of the network's previous state. This affects the dependent representations only. For example, with independent $\mathbf{C} \rightarrow \mathbf{Z}$ vectors the network correctly determines mixtures that do not include **A** or **B**. If **A** or **B** are present in the mixture, then there is 50% chance of being correct (depends on initial conditions which are randomized). Dependent label conflicts can also occur with combinations of representations. For example, if $\mathbf{C} + \mathbf{D} = \mathbf{E}$ (Case II) then activation of **E** will be distributed to nodes y_C, y_D, and y_E. The network's responses to dependent representations are logical because the network does not have a way to differentiate between the dependent representations. However, learning algorithms should strive to avoid linear dependent representations so that recognition can be unambiguous. One approach to avoiding linear de-

	Case I		Case II			Case III	
	A	B	C	D	E	A	A
x_1	0	0	1	0	1	1	1
x_2	1	1	1	0	1	1	1
x_3	0	0	0	0	0	0	1
$x_{..}$	0	0	0	0	0	0	0
x_i	1	1	0	1	1	1	0

Fig. 4.11 Types of degenerate training cases. Case I: same vectors with different labels. Case II: Two vectors equivalent to a third. Case III: Different vectors with the same label, commonly presented in a traditional learning scenario

pendence is to introduce hierarchical layers to the network whenever independence is violated [3].

However, if recognition information is sufficiently impoverished or incomplete, and only reveals features common to two patterns, or is an illusion such as Fig. 4.5, this can also place the network in the linear dependence state. The network cannot decide between multiple equal solutions. In such cases, network response depends on initial conditions: a semi-plastic regime that maintains memory of the network's previous state. This is a rational strategy to address incomplete information.

4.5.1.1 Plasticity and Neuroscience Implications

Neuroscience experiments that are commonly interpreted as a test for memory place feedback networks in the linear dependence regime. This semi-plastic regime is also a feedback alternative to the synaptic plasticity hypothesis. Commonly neuroscience experiments are cited as evidence of parameter optimization through the idea of Hebbian synaptic plasticity. Synaptic plasticity in this context of learning is the most studied phenomenon in neuroscience. Synaptic plasticity is assumed to occur in sensory regions, whenever a long-lasting change in communication between two neurons occurs as a consequence of stimulating them simultaneously. However, the mechanisms that control activity-dependent synaptic plasticity remain unclear because such experiments are variable [5, 17]. Furthermore, regulatory feedback networks can emulate phenomena seen in experiments (without synaptic plasticity), suggesting an alternate explanation for such learning [2].

Clearly, robust learning occurs in the brain. However the underlying mechanisms for plasticity (e.g., Spike timing dependent plasticity) may involve dynamics.

4.5.1.2 Non-binary Connection Weights

The focus of this chapter is to demonstrate the dynamics of this network and its ability to recognize combinations of patterns. For simplicity and to better demonstrate the plasticity mechanism due to dynamics, I focused on binary connections and learning the connectivity was trivial. Connectivity can be extended to weighted connections, however this is outside the scope of the chapter.

4.5.2 Measuring Difficulty Processing

The difficulty of mixture pattern processing can be evaluated by the number of cycles required to reach steady state. The greater number of cycles required the harder the task. The number of iterations needed increased linearly with the number of patterns mixed. It took an average of 23 cycles in the 1 pattern case, 42 cycles in the $k = 2$ mixture pattern case, and 82 cycles in the $k = 4$ pattern case to reach steady state. It is important to note that the difficulty of this task as measured by cycle time increases only linearly with the number of combinations even while the number of combinations increase exponentially.

4.5.2.1 Notes on Cycles

The number of iterations depends on the definition of steady state and the value of dt utilized in the equation. The number of iterations necessary can vary based on a speed accuracy trade-off. To be very accurate, the criteria for steady state (and dt) should be small. dt remained 1 throughout all simulations and steady state was declared when the sum of y's do not vary more than 0.0001 between cycles. To increase speed, the criteria for steady state can be increased.

4.5.3 Combined Oscillatory and Non-oscillatory Dynamics

Like feedforward methods, this method does not preclude additional oscillatory units to enhance processing, and future studies in combining these approaches are planned.

4.5.4 Summary

This work suggests that (1) the brain may not perform recognition via feedforward models (2) Self-Regulatory Feedback is a major contributor to flexible processing and binding. The feedback perspective may lead to a better understanding of animals' ability to efficiently interpret and segment scenes.

Acknowledgements I would like to thank Eyal Amir, Robert Lee DeVille, Dervis Vural, Ravi Rao, and Cyrus Omar for discussion and suggestions; Cyrus Omar also assisted with simulation software for testing. This work was in part supported by the U.S. National Geospatial Agency and IC Postdoc Grant HM1582-06–BAA-0001.

References

1. Achler T (2007) Object classification with recurrent feedback neural networks. In: Evolutionary and bio-inspired computation: theory and applications. Proc SPIE, vol 6563

2. Achler T (2008) Plasticity without the synapse: a non-Hebbian account of LTP. Society for Neuroscience poster, Washington
3. Achler T (2009) Using non-oscillatory dynamics to disambiguate simultaneous patterns. In: Proceedings of the 2009 IEEE international joint conference on neural networks (IJCNN'09)
4. Achler T, Amir E (2008) Input feedback networks: classification and inference based on network structure. J Artif Gen Intell 1:15–26
5. Achler T, Amir E (2009) Neuroscience and AI share the same elegant mathematical trap. J Artif Gen Intell 2:198–199
6. Achler T, Omar C et al (2008) Shedding weights: more with less. In: Proceedings of the 2008 IEEE international joint conference on neural networks (IJCNN'08), pp 3020–3027
7. Achler T, Vural D et al (2009) Counting objects with biologically inspired regulatory-feedback networks. In: Proceedings of the 2009 IEEE international joint conference on neural networks (IJCNN'09)
8. Ackerman S. (2010) $19 billion later, Pentagon's best bomb-detector is a dog. Wired Magazine, http://www.wired.com/dangerroom/2010/10/19-billion-later-pentagon-best-bomb-detectoris-a-dog/
9. Aroniadou-Anderjaska V, Zhou FM et al (2000) Tonic and synaptically evoked presynaptic inhibition of sensory input to the rat olfactory bulb via GABA(B) heteroreceptors. J Neurophysiol 84(3):1194–1203
10. Atema J (1988) Distribution of chemical stimuli. In: Sensory biology of aquatic animals. Springer, New York, pp 29–56, xxxvi, 936 p
11. Bab-Hadiashar A, Suter D (2000) Data segmentation and model selection for computer vision: a statistical approach. Springer, New York
12. Carpenter GA, Grossberg S (1987) A massively parallel architecture for a self-organizing neural pattern-recognition machine. Comput Vis Graph Image Process 37(1):54–115
13. Chen WR, Xiong W et al (2000) Analysis of relations between NMDA receptors and GABA release at olfactory bulb reciprocal synapses. Neuron 25(3):625–633
14. Douglas RJ, Martin KA (2004) Neuronal circuits of the neocortex. Annu Rev Neurosci 27:419–451
15. Famiglietti EV Jr, Peters A (1972) The synaptic glomerulus and the intrinsic neuron in the dorsal lateral geniculate nucleus of the cat. J Comp Neurol 144(3):285–334
16. Felleman DJ, Van Essen DC (1991) Distributed hierarchical processing in the primate cerebral cortex. Cereb Cortex 1(1):1–47
17. Froemke RC, Tsay IA et al (2006) Contribution of individual spikes in burst-induced long-term synaptic modification. J Neurophysiol 95(3):1620–1629
18. Garthwaite J, Boulton CL (1995) Nitric oxide signaling in the central nervous system. Annu Rev Physiol 57:683–706
19. Herd SA, O'Reilly RC (2005) Serial visual search from a parallel model. Vis Res 45(24):2987–2992
20. Itti L, Koch C (2001) Computational modelling of visual attention. Nat Rev Neurosci 2(3):194–203
21. Koch C, Ullman S (1985) Shifts in selective visual attention: towards the underlying neural circuitry. Hum Neurobiol 4(4):219–227
22. LaBerge D (1997) Attention, awareness, and the triangular circuit. Conscious Cogn 6(2–3):149–181
23. Landisman CE, Connors BW (2007) VPM and PoM nuclei of the rat somatosensory thalamus: intrinsic neuronal properties and corticothalamic feedback. Cereb Cortex 17(12):2853–2865
24. Latecki L (1998) Discrete representation of spatial objects in computer vision. Kluwer Academic, Dordrecht
25. Marder E, Goaillard JM (2006) Variability, compensation and homeostasis in neuron and network function. Nat Rev Neurosci 7(7):563–574
26. Mcfadden FE (1995) Convergence of competitive activation models based on virtual lateral inhibition. Neural Netw 8(6):865–875
27. Mozer MC (1991) The perception of multiple objects: a connectionist approach. MIT Press, Cambridge

28. Navalpakkam V, Itti L (2007) Search goal tunes visual features optimally. Neuron 53(4):605–617
29. Pearce TC (2003) Handbook of machine olfaction: electronic nose technology. Wiley-VCH, Weinheim
30. Rachkovskij DA, Kussul EM (2001) Binding and normalization of binary sparse distributed representations by context-dependent thinning. Neural Comput 13(2):411–452
31. Rao AR, Cecchi GA et al (2008) Unsupervised segmentation with dynamical units. IEEE Trans Neural Netw 19(1):168–182
32. Rosenblatt F (1962) Principles of neurodynamics; perceptrons and the theory of brain mechanisms. Spartan Books, Washington
33. Treisman A, Gormican S (1988) Feature analysis in early vision—evidence from search asymmetries. Psychol Rev 95(1):15–48
34. Treisman AM, Gelade G (1980) A feature-integration theory of attention. Cogn Psychol 12(1):97–136
35. Turrigiano GG, Nelson SB (2004) Homeostatic plasticity in the developing nervous system. Nat Rev Neurosci 5(2):97–107
36. von der Malsburg C (1999) The what and why of binding: the modeler's perspective. Neuron 24(1):95–104, 111–125
37. Witten IH, Frank E (2005) Data mining: practical machine learning tools and techniques. Morgan Kaufmann, Amsterdam
38. Wolfe JM (2001) Asymmetries in visual search: an introduction. Percept Psychophys 63(3):381–389

Chapter 5
Functional Constraints on Network Topology via Generalized Sparse Representations

A. Ravishankar Rao and Guillermo A. Cecchi

Abstract A central issue in neuroscience is understanding to what extent the observed neural network architectural patterns are the result of functional constraints, as opposed to evolutionary accidents. In this regard, the visual cortex exhibits interesting regularities such that the receptive field sizes for feed-forward, lateral and feedback connections are monotonically increasing as a function of hierarchy depth. In this chapter, we use computational modeling to understand the behavior of the visual system and propose an explanation of this observed connectivity pattern via functional metrics based on separation and segmentation accuracies. We use an optimization function based on sparse spatio-temporal encoding and faithfulness of representation to derive the dynamical behavior of a multi-layer network of oscillatory units. The network behavior can be quantified in terms of its ability to separate and segment mixtures of inputs. We vary the topological connectivity between different layers of the simulated visual network and study its effect on performance in terms of separation and segmentation accuracy. We demonstrate that the best performance occurs when the topology of the simulated system is similar to that in the primate visual cortex where the receptive field sizes of feedforward, lateral and feedback connections are monotonically increasing.

5.1 Introduction

Connections between neurons throughout the brain, and particularly in the cortex, create complex networks which are both hierarchically organized and also contain multiple parallel pathways [1]. At the same time, there is clear evidence that the cortical architecture is not univocally determined, and that moreover it is highly plastic, changing in the scale of milliseconds to years. An essential question, therefore, is to understand how neural circuits utilize these different connections, and how the dynamics of signal propagation through these networks relate to function [2]; in

A.R. Rao (✉) · G.A. Cecchi
IBM Research, Yorktown Heights, NY 10598, USA
e-mail: ravirao@us.ibm.com

G.A. Cecchi
e-mail: gcecchi@us.ibm.com

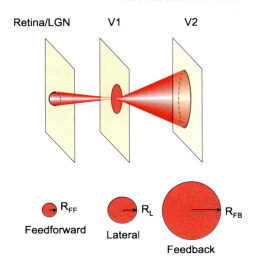

Fig. 5.1 The relative projection radii for the different classes of cortical connections that terminate in a neuron in V1. The projection radius is the radius of neurons which project to a target V1 neuron. The feedforward projection radius, is R_{FF}, the lateral projection radius is R_L and the feedback projection radius is R_{FB}. This figure is adapted from Angelucci and Bullier [9]

other words, to understand to what extent, if at all, function imposes constraints on topology.

The National Academy of Engineering in 2008 identified the reverse-engineering of the brain as a grand challenge (http://www.engineeringchallenges.org). Due to the complexity of the problem, insights will be required from multiple disciplines. Sporns et al. [3] recommend creating a detailed structural model that captures the connectivity amongst the neurons in the brain. This model would be useful in performing detailed neural simulations and in understanding the propagation of signals in the network. Kasthuri and Lichtman [4] view the connectivity map as providing an anatomical basis to understand information processing within a brain. Seth et al. [5] describe a computational model of brain function built through the knowledge of anatomical connectivity.

Computational modeling plays an essential role in understanding the relationship between neural structure and organism behavior [6], as exemplified by Shepherd et al. [7]. In spite of much progress in this area, there are several open questions; among these, and for reasons related to the availability of experimental observations and the relative ease of functional interpretations, we will focus on the following: why do different types of connectivity patterns exist for feedforward, lateral and feedback connections in the visual cortex? An answer to this question will have vast implications, ranging from computer vision to basic neuroscience. Indeed, as the editors of the experimentally-oriented Journal of Physiology observe, "The discovery of the highly complicated structure of visual (but also nonvisual) cortical areas, including their horizontal and vertical organization and their intrinsic diversity, calls for new conceptual tools that appear necessary to handle their architectural and functional aspects" [8].

The current understanding of the pattern of connectivity in the visual cortex is summarized in Fig. 5.1 [9]. A striking feature of this pattern is the following relationship between the feedforward projection radius R_{FF}, lateral projection radius, R_L, and feedback projection radius, R_{FB}:

$$R_{FF} < R_L < R_{FB}; \quad R_L \sim 2R_{FF}; \quad R_{FB} \sim 2R_L. \tag{5.1}$$

Such a clear relationship begs the following questions: why does this pattern of connectivity exist, and what is its role? The main contribution of this chapter is to present an explanation of the functional constraints that may give rise to this topological regularity, using the framework of an optimization approach for object representation in the visual cortex. We use a hierarchical network of oscillatory elements, where the different connection weights are learnt through unsupervised learning. We present a detailed exploration of the effect of topological connectivity on network behavior. We use the quantitative measures of separation and segmentation accuracies to capture the functioning of the network. With these tools, we show that structure and function questions can be answered by establishing quantitative metrics on the performance of the neural system. The behavior of the system can be understood via dynamics and parameter selections that optimize the chosen metrics. Using this approach, we show that the observed topological connectivity in the cortex establishes an optimum operating point for a neural system that maximizes the classification accuracy for superposed visual inputs, and their corresponding segmentation.

Our problem formulation and network structure is extensible to multiple levels of a hierarchy, and this should spur further effort in analyzing more levels in the visual system. The size of the network can also be increased, in terms of the number of units, leading to more detailed modeling of visual object representation and processing.

5.2 Background

There are two perspectives that we bring together in this chapter, involving function and structure within the cortex. We first examine the issue of function. Let us consider the development of the human visual system (HVS) and identify a few of the key problems it must solve: (1) It should create high-level object categories in an unsupervised manner [10]. (2) It should separate mixtures of inputs into their constituents. (3) It should establish a correspondence between lower level image features such as pixels or edges and the higher level objects that they constitute. This process is termed *segmentation*, which groups high level percepts and supporting low level features. (4) It should handle different types of signal degradation, involving noise and occlusion. This may require the use of multi-layer systems and redundancy, as seen in the visual pathway.

How does the visual system achieve these objectives? Some clues are provided by examining the time-domain behavior of neurons. Ensembles of neurons exhibit interesting time-domain behavior such as oscillation and synchronization [11]. The computational modeling of oscillatory networks has been studied by numerous researchers [12, 13]. A desirable property achieved through synchronization is that of segmentation, which refers to the ability of a network to identify the elements of the input space that uniquely contribute to each specific object. For visual inputs, segmentation would establish a correspondence between the pixels or edges and the higher-level objects they belong to. Segmentation can also be considered to be a solution to the binding problem [14], and could lead to advances in neural networks for object recognition. Rao et al. [15] developed a computational model that uses synchronization to perform segmentation of visual objects. This model was initially developed for two-layer networks, but has been extended for multi-layer networks [16, 17]. We briefly review and refine this model over the next two sections.

We now examine our second perspective, covering cortical structure. Felleman and van Essen [1] created a computerized database that stored known connectivity information between different cortical areas. Their analysis of this data led to the observation that the cortex is hierarchically organized with multiple intertwined processing streams. In addition to the hierarchical pathways, there are multiple parallel pathways, which make the cortical system highly distributed. Furthermore, a majority of inter-areal connection pathways are reciprocal. Their analysis of connectivity patterns was based on the origin and termination of connections within the six layered laminar cortical structure. Feedforward connections originate in the superficial layers, from layers 1–3, and terminate mainly in layer 4. In contrast, feedback connections originate in the infragranular layers, from layers 5–6 and terminate outside layer 4.

It has been estimated that the number of feedback connections in the macaque monkey going from V2 to V1 is roughly ten times the number of feedforward geniculocortical afferents [18]. Furthermore, the number of feedforward geniculate synapses found in V1 is smaller than the number of synapses formed with other cells in V1 [18]. It has been suggested that the feedback connections are diffuse, but sparse, as a single feedback neuron is unlikely to project to all neurons in V1 within its receptive field [19]. Thus, feedback connections are not simply the inverse of feedforward connections. Furthermore, the overall effect of feedback connections is weak, and not sufficient to sustain activity in V1 neurons in the absence of geniculocortical input.

Studies have shown that feedback inputs usually add to the feedforward inputs to a cortical area, thereby increasing the activity of the target neurons [21]. Furthermore, lateral connections created by horizontal axons do not drive their target neurons like feedforward geniculo-cortical inputs, but rather cause modulatory subthreshold effects [20]. Though early studies hinted that feedback connections are diffuse, evidence is emerging that feedback may be highly patterned [21].

We bring together the two perspectives on function and structure by varying the connectivity of the network in our model, and observing its effect on network function. We specifically examine the function of the network as it relates to the binding problem, and observe how well it is able to perform the segmentation of superposed inputs. Though this is currently an empirical approach, it provides an insight into the behavior of an oscillatory network that achieves object recognition.

5.3 Network Dynamics and Learning

We have proposed an objective function that governs the network dynamics and learning in multi-layered networks [17]. This approach first examines the network dynamics in a two-layered system consisting of oscillatory elements.

5.3.1 An Objective Function for Network Performance

Consider a network with inputs **x** and output **y**. Let **x** and **y** be represented as row vectors, and let **W** be a matrix representing synaptic weights that transform **x** into **y**. The outputs y_i are constrained to be nonnegative, that is, $y_i \geq 0$ $\forall i$. Note that the output $\mathbf{y} \neq \mathbf{W}\mathbf{x}^T$ due to the presence of lateral and feedback interactions, and the nonlinear constraint to make **y** nonnegative.

We use the principle of sparse representation (cf. [22]) to derive an objective function E, as follows. The underlying biological interpretation is that cortical processing creates a sparse representation of its input space.

$$E = \left\langle \mathbf{y}^T \mathbf{W} \mathbf{x} - \frac{1}{2}\mathbf{y}^T \mathbf{y} - \frac{1}{2}\sum_n \mathbf{W}_n^T \mathbf{W}_n + \frac{1}{2}\lambda \mathrm{S}(\mathbf{y}) \right\rangle_\mathcal{E}, \tag{5.2}$$

where $\mathcal{E}$ represents the input ensemble. The first term captures the faithfulness of representation, and rewards the alignment between the network's output and the feed-forward input. The second term is a constraint on the global activity. The third term is achieves normalization of the synaptic weight vectors. The last term consists of the variance of the outputs, and rewards the sparseness of the representation:

$$\mathrm{S}(\mathbf{y}) = N\left(\langle y_n^2 \rangle_\mathcal{N} - \langle y_n \rangle_\mathcal{N}^2\right) = \sum_{n=1}^{N} y_n^2 - \frac{1}{N}\left(\sum_{n=1}^{N} y_n\right)^2. \tag{5.3}$$

By maximizing the objective function *wrt* **y**, one obtains the dynamics that maximizes it upon presentation of each input, and maximizing it *wrt* **W** one obtains the optimal learning update. The maximization can be achieved through gradient ascent.

The instantaneous state of the amplitude-based network described above corresponds to a vector in real space $\mathbb{R}^N$. For oscillatory network units, whose state is determined by an amplitude and a phase, we can represent the activity of each unit by a vector in complex space $\mathbb{C}^N$. This corresponds to a phasor notation of the activity in the form $x_k e^{i\phi_k}$ for the lower layer and $y_k e^{i\theta_k}$ for the upper layer, where ϕ_k and θ_k are the phases of the kth unit in the lower layer and upper layer, respectively.

This allows us to generalize the objective function in Eq. 5.2 through the quantity E_s, where

$$E_s = E + \beta \mathrm{Re}\big[\mathcal{C}(E)\big]. \tag{5.4}$$

Here, $\mathcal{C}[E] = E(\mathbf{p}, \mathbf{q})$ is the complex extension of the energy

$$\mathcal{C}(E) = \mathbf{q}^\dagger \mathbf{W} \mathbf{p} + \frac{1}{2}\lambda \mathrm{S}(\mathbf{q}) - \frac{1}{2}\mathbf{q}^\dagger \mathbf{q} \tag{5.5}$$

such that $p_n = x_n e^{i\phi_n}$, $q_n = y_n e^{i\theta_n}$, $\dagger$ is the transpose conjugate operator, and $\mathsf{S}(\mathbf{q})$ is the complex extension of the variance:

$$\mathsf{S}(\mathbf{q}) = \sum_{n=1}^{N} q_n^\dagger q_n - \frac{1}{N}\left(\sum_{n=1}^{N} q_n^\dagger\right)\left(\sum_{n=1}^{N} q_n\right). \tag{5.6}$$

In these equations, the parameter β determines the weight given to the oscillatory part of the energy, where the case $\beta = 0$ represents a traditional neural network without oscillatory units. The maximization of E_s leads to an efficient segmentation of the inputs.

5.3.2 Extension to Multi-layered Networks

We extend the objective function defined in Eq. 5.4 to an array of M layers connected sequentially as follows. Let i and j refer to two sequential layers. The global objective function for a multi-layered system with M layers is defined by

$$E_S^G = \sum_{i=0}^{M-1} E_S(x_i, x_{i+1}), \tag{5.7}$$

where G denotes the global objective function, and x_0 is the input layer. The rationale used here is that successive pairs of layers are governed by the local optimization in Eq. 5.4.

In this chapter, we consider a three-layered network consisting of a lower input layer $\{x\}$, a middle layer $\{y\}$ and an upper layer $\{z\}$. The corresponding objective function is

$$E_S^G = E_S(x, y) + E_S(y, z). \tag{5.8}$$

Let x, y and z denote the amplitudes of units in lower, middle and upper layers. Similarly, the phases are represented by ϕ in the lower layer, θ in the middle layer, and ξ in the upper layer.

5.4 Network Configuration

Each oscillating unit is described by its frequency, amplitude, phase. A unit's phase is derived from an ongoing oscillation with a fixed natural period. A unit's period is randomly drawn from the range $\tau \in [2.0, 2.1]$ msec. The network dynamics obtained through the maximization of the objective function determine the evolution of the amplitudes and phases.

Figure 5.2 shows the network. Rather than using all-to-all connections, we employ restricted connectivity between the layers, which *qualitatively* models the topological connectivity patterns observed in primate cortex. Let N_x^{FF} be the neighbor-

5 Functional Constraints on Network Topology

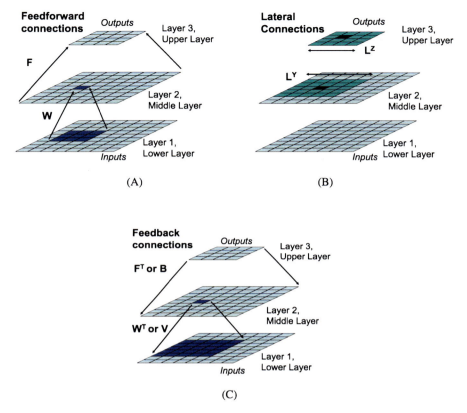

Fig. 5.2 A depiction of the network connectivity. There are three layers, whose units are denoted by {x} for layer 1, the lower input layer; {y} for layer 2, the middle layer; and {z} for layer 3, the upper layer. (**A**) Feedforward connections. (**B**) Lateral connections. (**C**) Feedback connections. Each unit in the upper layer is connected to all units in the middle layer

hood in the input layer that contributes to the activity of the nth unit of the middle layer. Similarly, let N_y^L be the lateral neighborhood in the middle layer that is connected to the nth unit of the middle layer, and let N_z^{FB} be the neighborhood in the upper layer that is connected via feedback to the nth unit of the middle layer.

5.4.1 Network Dynamics

Let **W** denote the connection weights from elements in {x} to {y}, **V** denote connections weights from {y} to {x}, **F** denote connections from {y} to {z} and **B** denote connections from {z} to {y}. Lateral connections in the {y} layer are denoted by $\mathbf{L}^Y$ and in the {z} layer by $\mathbf{L}^Z$.

The amplitude and phase for the middle layer units evolve as follows:

$$\Delta y_n \sim \sum_{j \in N_x^{FF}} W_{nj} x_j \left[1 + \beta \cos(\phi_j - \theta_n)\right]$$
$$+ \kappa_1 \sum_{k \in N_z^{FB}} B_{nk} z_k \left[1 + \beta \cos(\xi_k - \theta_n)\right]$$
$$- \alpha y_n - \gamma \sum_{m \in N_y^L} L_{nm}^Y y_m \left[1 + \beta \cos(\theta_m - \theta_n)\right],$$
$$\Delta \theta_n \sim \beta \sum_{j \in N_x^{FF}} W_{nj} x_j \sin(\phi_j - \theta_n)$$
$$+ \kappa_2 \beta \gamma \sum_{m \in N_z^{FB}} B_{nm} z_m \sin(\xi_m - \theta_n)$$
$$- \beta \gamma \sum_{k \in N_y^L} L_{nk}^Y y_k \sin(\theta_k - \theta_n),$$
(5.9)

where κ_1 and κ_2 are weighting terms for the relative contribution of the feedback inputs.

For the upper layer, we obtain:

$$\Delta z_n \sim \sum_{j \in N_y^{FF}} F_{nj} y_j \left[1 + \beta \cos(\theta_j - \xi_n)\right] - \alpha z_n$$
$$- \gamma \sum_{k \in N_z^L} L_{nk}^Z z_k \left[1 + \beta \cos(\xi_k - \xi_n)\right],$$
$$\Delta \xi_n \sim \beta \sum_{j \in N_y^{FF}} F_{nj} y_j \sin(\theta_j - \xi_n)$$
$$- \beta \gamma \sum_{k \in N_z^L} L_{nk}^Z z_k \sin(\xi_k - \xi_n).$$
(5.10)

In this equation, N_y^{FF} denotes the neighborhood in the middle layer that provides feed-forward connections to the nth unit of the upper layer. Similarly, N_z^L is the lateral neighborhood in the upper layer that connects to the nth unit of the upper layer.

In the lower layer, only the phase is a free variable, as the amplitude is determined by the input presented to the system. The lower layer phase varies as follows:

$$\Delta \phi_n \sim \sum_{j \in N_y^{FB}} V_{nj} y_j \sin(\theta_j - \phi_n), \qquad (5.11)$$

where N_y^{FB} is the neighborhood in the middle layer that is connected via feedback to the nth unit of the lower layer.

We use a phase-dependent Hebbian learning rule to modify synaptic weights R_{ij} between a source node with index j and a target node with index i as follows

$$\Delta R_{ij} \sim S_j T_i \left[1 + \beta \cos(\Delta)\right]. \qquad (5.12)$$

5   Functional Constraints on Network Topology   83

Here, S_j is amplitude of the source node, T_i is the amplitude of the target node, and Δ is the phase difference between the source and target node. The source and target nodes can be picked from any two sequentially connected layers. We also permit lateral weights in the $\{y\}$ and $\{z\}$ layers to learn.

5.4.2 Computational Modeling: Inputs and Network Behavior

Our system operates on inputs consisting of 2D gray level images. We use a set of 16 objects shown in Fig. 2 in [15]. Some examples from this set are shown in Figs. 5.3 and 5.4.

The network operates in two sequential stages: learning and performance. During the first stage consisting of unsupervised learning, elements of the input ensemble are presented to the network. The response of the network is dynamically computed according to Eqs. 5.9 through 5.11. We observe that the amplitude activity in the network reaches a steady state after approximately 300 iterations. The learning rule given by Eq. 5.12 is applied only after the network amplitude activity has settled down. This process prevents the network from learning associations during the early settling period. This overall process is repeated for 1,000 presentations. We choose a nominal settling period of $T = 300$ iterations, or 7.5 cycles of oscillation. The typical behavior of the system is that a single unit in the output layer emerges as a winner for a given input. Furthermore, after the 1,000 trials, a unique winner is associated with each input.

5.5 Dynamical Segmentation

The second stage, called the performance stage, consists of an evaluation of the system performance when it is presented with single objects or a superposition of two objects. The system described in Sect. 5.4 is presented with a randomly chosen image from the input ensemble. We observe a winner-take-all dynamics upon presentation of one of the learned inputs. When the system is presented with a superposition of two randomly selected input objects, the output layer contains two winners. These winners tend to synchronize with units in the middle and lower layers as described in [15]. This tests the ability of the network to separate these superposed inputs. Ideally, we'd like to see the phase of a winner in the upper layer synchronized with object features in the lower layers that describe the particular object that the winner has learnt to represent.

We use Fig. 5.3 to explain the phase behavior of the system. Each unit in the oscillatory network is represented by a phasor, which is a vector with an amplitude and a phase. The 2-D display of phasors can be considered to be a vector field. The input to the system consists of a superposition of Objects 13 and 4. The vector fields displayed here demonstrate that most units in the middle and lower layers are synchronized with the two winners in the upper layer. Furthermore, the units that have similar phase in the lower layer units tend to represent a single object. This can be seen from the silhouettes in the vector field of lower layer activity. In order to clarify this phenomenon of segmentation, we display the segmented lower layer according to phase. The bottom two panels of Fig. 5.3 show that lower layer units are synchronized with each of the winners. Similarly, Fig. 5.4 shows the network activity for a superposition of objects 12 and 3.

These results imply that the phase information can be used to convey relationship information between different layers in a hierarchy, which is not possible if only amplitude information is used. Thus, if an action needs to be taken based on the identification of a certain object at a higher layer, the phase information provides information about where that object is localized in the lower layer. One can simply use the phase information across multiple layers as an index. This is the essence of the binding problem as explained in Sect. 5.1, as it is the phase information that "binds" together the amplitude information in multiple layers and establishes relationships.

In order to quantify the behavior of the network, we measure two aspects of the system response, $\mathbf{z}$. The first aspect determines whether the winners for the superposed inputs are related to the winners when the inputs are presented separately. We term this measurement the *separation accuracy*, defined as follows. Let unit i in the upper layer be the winner for an input $\mathbf{x}_1$, and let unit j be the winner for input $\mathbf{x}_2$. If units i and j in the upper layer are also winners when the input presented is $\mathbf{x}_1 + \mathbf{x}_2$, then we say the separation is performed correctly, otherwise not. The ratio of the total number of correctly separated cases to the total number of cases is defined to be the separation accuracy.

The second aspect determines the accuracy of phase segmentation, which corresponds to how well the phases of the input layer units are synchronized with their higher-level categorizations in the output layer. Phase segmentation accuracy can be measured by computing the fraction of the units of the lower layer that correspond to a given object and lie within some tolerance of the phase of the upper layer unit that represents the same object.

We measure the separation and segmentation accuracy over three independent sets of 50 trials using pairs of randomly selected inputs.

We investigated the relationship between structure and function by varying the connectivity parameters between and within the different layers and measuring their effect on the function of the network. This is described in Sect. 5.6. We used the following model parameters: $\beta = 0.5$, $\alpha = 0.25$, $\gamma = 0.02$, $\kappa_1 = 0.003$, $\kappa_2 = 1$. This method provides a quantitative basis for the measurement of function.

5.5.1 Alternate Methods of Measuring System Performance

The measurement of global synchrony has been proposed in the literature as a way of measuring the behavior of networks of oscillatory elements [23]. This has proven to be a useful measure in investigating the relationship between connectivity and

5 Functional Constraints on Network Topology

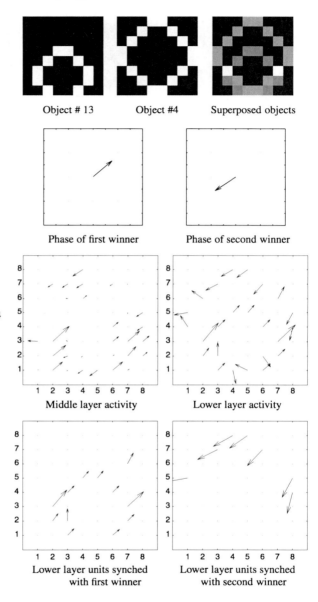

Fig. 5.3 An illustration of the behavior of phase information. This figure shows the superposition of objects 13 and 4. The phase of the first winner corresponding to object 13 in the upper layer is 0.71 whereas the phase of the second winner in the upper layer corresponding to object 4 is 3.73. The activity in the middle and lower layer units is displayed as a 2D vector field. The magnitude of the vector reflects the amount of activity in the unit, and the direction encodes the phase of the unit. Note that the silhouette of the lower layer units synchronized with the first winner resembles object 13. A similar observation holds for object 4

the ability of elements in a network to synchronize. For instance, Osipov et al. [23] showed that networks with more heterogeneous degrees generate weaker collective synchronization. They also examined the effect on synchronization of introducing short-cuts in networks such that it resulted in the creation of small-world networks.

One of the measures suggested for global synchrony is based on variance as follows. Let $x_i(t)$ denote the activity of the ith neuron in an ensemble of N neurons

Fig. 5.4 An illustration of the behavior of phase information. This figure shows the superposition of objects 12 and 3. The phase of the first winner corresponding to object 12 in the upper layer is 0.505, whereas the phase of the second winner in the upper layer corresponding to object 3 is 3.98. Note that the silhouette of the lower layer units synchronized with the first winner resembles object 12. A similar observation holds for object 2

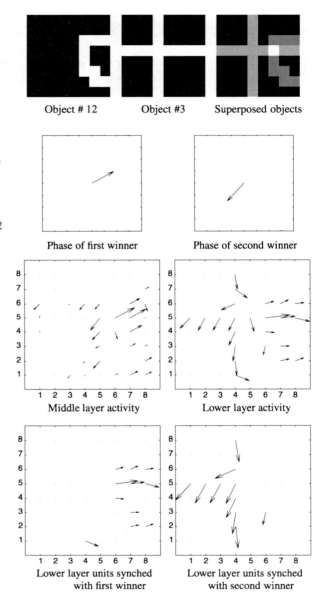

measured at different times, t. The average activity at a given time t is given by $X(t)$ as follows

$$X(t) = \frac{1}{N} \sum_{i=1}^{N} x_i(t). \tag{5.13}$$

5 Functional Constraints on Network Topology

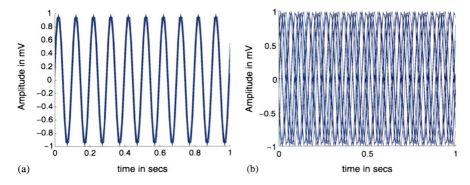

Fig. 5.5 Measurement of global synchrony in a network of 10 simulated oscillator units. (**a**) Shows the amplitude trace of the units with small phase differences. The global synchrony measure S was computed to be 0.969. (**b**) Shows the amplitude trace of the units with larger phase differences. The global synchrony measure S was computed to be 0.154

The variance of the quantity $X(t)$ is calculated over the discrete times over which the measurements are made.

$$\sigma_X^2 = \langle [X(t)]^2 \rangle_t - [\langle X(t) \rangle_t]^2, \tag{5.14}$$

where the operator $\langle \ldots \rangle_t$ denotes the expected value or a time averaging over a sufficiently large time window. We also calculate the variance of the individual neuronal activities as follows

$$\sigma_{x_i}^2 = \langle [x_i(t)]^2 \rangle_t - [\langle x_i(t) \rangle_t]^2. \tag{5.15}$$

The synchrony measure, S is defined as follows

$$S = \frac{\sigma_X^2}{\frac{1}{N}\sum_{i=1}^{N}\sigma_{x_i}^2}. \tag{5.16}$$

This measure is between 0 and 1, with 1 indicating perfect synchrony. Perfect synchrony occurs when the $x_i(t)$ are identical.

To illustrate this, we apply the method as follows to an ensemble consisting of 10 oscillating neurons. For the sake of simplicity, the activity of each neuron is modeled by the following equation

$$x_i(t) = \cos(2\pi f t + \phi_i), \tag{5.17}$$

where f is the frequency, and ϕ_i is the phase. We set $f = 10$ Hz. The phase ϕ_i is randomized in two different ways as follows. Let r be a random number between 0 and 1 chosen from a uniform distribution. In the first case, we use $\phi_i = 2\pi r/10$. This results in small phase differences between the signals $x_i(t)$. In the second case, we use $\phi_i = 2\pi r$, which results in larger phase differences. Figure 5.5 illustrates these two cases. The synchrony measure computed was $S = 0.969$ in the first case, and $S = 0.154$ in the second.

We now apply the synchrony measure S, as defined in Eq. 5.16 to the behavior of units in the multi-layer oscillatory network in our model. We apply the measure to units in each individual layer, namely Layer 2 and Layer 3 separately.

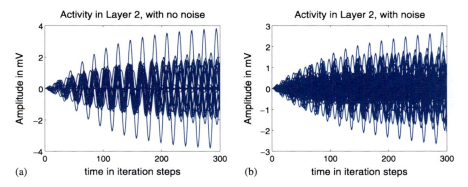

Fig. 5.6 (a) Shows the amplitude trace of the units in Layer 2 of the oscillatory network depicted in Fig. 5.2. The input consisted of a superposition of two objects, consisting of grayscale images from the collection in Fig. 2 in [15]. No input noise was used in this case. (b) Shows the amplitude trace of the units in Layer 2, when the input is perturbed with 10% additive noise

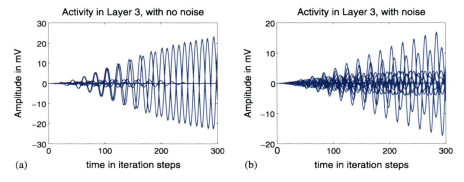

Fig. 5.7 (a) Shows the amplitude trace of the units in Layer 3 of the oscillatory network depicted in Fig. 5.2. The input consisted of a superposition of two objects, consisting of grayscale images from the collection in Fig. 2 in [15]. No input noise was used in this case. (b) Shows the amplitude trace of the units in Layer 3, when the input is perturbed with 10% additive noise

For the sake of illustration, in Fig. 5.6 we show the time course of activity in the units in Layer 2. We use two cases, one where the input is noise free, and another where the input consists of 10% additive noise drawn from a uniform distribution. Similarly, Fig. 5.7 shows the activity of units in Layer 3.

We conducted a set of 100 experiments as follows. In the kth experiment, we first start with an oscillatory network with randomized weights, and perform an unsupervised training as described in Sect. 5.3. For this network, we randomly select a pair of objects to be superposed, and compute the time course of network activity in all the units. For each time evolution, we compute the synchrony measure $S_{2,k}$ for Layer 2 and $S_{3,k}$ for Layer 3. In the subscript $(2, k)$ the first index denotes the layer over which the synchrony measure is computed, and the second index k describes the ordinal number of the experiment.

5 Functional Constraints on Network Topology 89

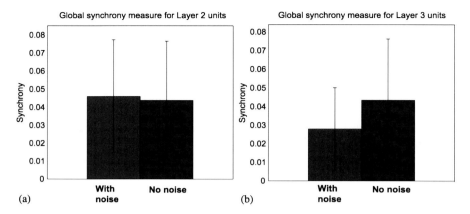

Fig. 5.8 Applying a global synchrony measure within each layer. (**a**) Shows the synchrony measure for Layer 2 units, with and without noisy inputs. (**b**) Shows the synchrony measure for Layer 3 units with and without noisy inputs. The *vertical lines* indicate error bars of plus or minus one standard deviation

We computed statistics for the distribution of $S_{2,k}$ and $S_{3,k}$, which are summarized in Fig. 5.8. From this result, we observe that there is no significant difference in the synchrony measure for each layer when the inputs are noisy or noise-less. Thus such a synchrony measure is not capable of distinguishing subtle network behaviors that occur when it is performing a specific task such as segmentation and separation of superposed inputs. More specific measures such as segmentation and separation accuracy as defined in Sect. 5.5 show that a significant performance degradation occurs in the presence of noise, as summarized later in Fig. 5.12.

5.6 Effect of Varying Geometric Connectivity

We explored the relationship between structure and function by performing a systematic study of the variation of network performance with different connection configurations. An advantage of our computational approach is that it provides objective measures of system performance relative to the task of visual object categorization and segmentation. We can then pose the question: what is the pattern of network connectivity that maximizes the system performance? The resulting pattern of connectivity is then compared with biological measurements to understand its significance. With this rationale, we explored three sources of variation in network connectivity arising from differing radii of feedforward, lateral and feedback projections, denoted by R_{FF}, R_L and R_{FB}. We were also motivated by the work of Przybyszewski [24], who summarized anatomical findings of massive back-projecting pathways in primate visual cortex. Though we have modeled only visual recognition, our method should be applicable to other sensory processing systems as well.

The space over which the results may be visualized is five dimensional, and consists of the variation of separation and segmentation accuracy over R_{FF}, R_{FB}

and R_L. To visualize this, we use a series of 3D plots in Fig. 5.9. We observe the following performance trends from these plots. As R_{FF} increases, the separation accuracy decreases, as we move from the plots (a) through (d). For a given value of R_{FF}, as R_{FB} increases, the separation accuracy increases. These trends are preserved for the segmentation accuracy plots, shown in Figs. 5.10(a)–(d).

Figures 5.11(a) and (b) show a series of 2D plots in order to provide a more detailed view of the data. For the chosen values of $R_{FF} = 1$, the observed trend is that with increasing R_L, separation and segmentation accuracies decrease. The best performance is obtained when the diameter of lateral interactions, $L = 3$, that is, $R_L = 1.5$, and at $R_{FF} = 0.5$ and $R_{FB} = 3.5$, which produces a separation accuracy of 98% and a segmentation accuracy of 97%.

We also examined the effect of noise on the system performance. We used uniformly distributed noise of 10% amplitude to distort the input image. The results are shown in Fig. 5.12. There is an overall decrease in separation accuracy when compared with Fig. 5.9. This decrease is particularly marked as R_{FF} is increased. The trends observed earlier also hold true in the presence of noise.

Figure 5.11 shows that the performance is also good when $R_L = 0.5$. However, the experiments conducted in the presence of noise show that performance degrades significantly for $R_L = 0.5$, as shown in Fig. 5.12(c). Hence, choosing $R_L = 1.5$ gives rise to the best overall performance.

These results show that the separation and segmentation accuracies serve as useful measures to understand the system performance as a function of precise network connectivity. From Fig. 5.11 we observe that when the diameter of lateral connectivity is 3, the separation accuracy is 98%, and segmentation accuracy is 97%, which is the best network performance for the given task. Our earlier approach presented in [25] achieved a separation of accuracy of 75% and a segmentation accuracy of 90%, as we did not explore the full range of connectivity possibilities.

5.7 Discussion

We have shown that an approach based on the objective function defined in Sect. 5.4 produces the following results:

1. It provides simple rules that govern neural dynamical and learning.
2. It creates a formulation that is extensible to multi-layer networks.
3. It achieves good computational results in terms of separating and segmenting superposed objects.
4. It provides a reasonable biological interpretation of segmentation, including the ability of the network to perform object categorization tasks in an unsupervised manner.
5. It provides an interpretation for the topology of feedforward, feedback and lateral connections in oscillatory networks that matches biological observations.

We offer the following qualitative interpretation of the relation between the different projection radii in Fig. 5.1. Small values of R_{FF} preserve the detail in the input

5 Functional Constraints on Network Topology

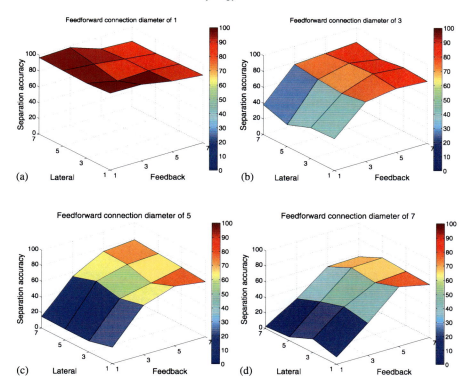

Fig. 5.9 Separation accuracy plotted as a function of network connectivity. (**a**) $R_{FF} = 0.5$ pixels for all cases, while R_{FB} and R_L vary as shown. (**b**) $R_{FF} = 1.5$ pixels. (**c**) $R_{FF} = 2.5$ pixels. (**d**) $R_{FF} = 3.5$ pixels. These plots show that separation accuracy decreases while R_{FF} increases, and while R_{FB} decreases. Reprinted with permission, from A.R. Rao, G.A. Cecchi, "An optimization approach to understanding structure-function relationships", International Joint Conference on Neural Networks, IJCNN 2009, pp. 2603–2610, © 2009 IEEE

images, and avoid blurring. Intermediate values of R_L are useful in combating input noise, and provide a smoothing capability. Larger values of R_{FB} allow higher layer categorizations to be shared with several units in the lower layer, thus providing top-down refinement of responses at lower layers. This framework not only makes intuitive sense: using a purely computational approach, Ullman et al. [26] showed that intermediate-complexity features are better suited for visual classification than very local or very global features. These intermediate feature carry more information about the class identity of objects such as cars, building or faces: it is better to have access to the presence of an eye and a nose, than the too general features of color- or orientation-dominated local patches, or the too specific complete picture. Our results agree with this observation, since the topological connectivity for optimal performance is neither one-to-one nor all-to-all. Moreover, the synchronization mechanism is designed, precisely, to *bind* features at different levels without collapsing them into a single scale of resolution.

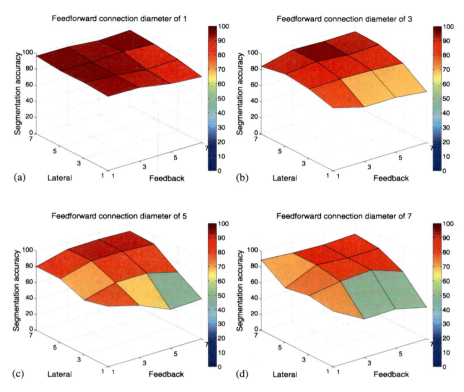

Fig. 5.10 Segmentation accuracy plotted as a function of network connectivity. (**a**) $R_{FF} = 0.5$ pixels for all cases, while R_{FB} and R_L vary as shown. (**b**) $R_{FF} = 1.5$ pixels for all cases. (**c**) $R_{FF} = 2.5$ pixels for all cases. (**d**) $R_{FF} = 3.5$ pixels for all cases. These plots show that segmentation accuracy decreases while R_{FF} increases, and while R_{FB} decreases. Reprinted with permission, from A.R. Rao, G.A. Cecchi, "An optimization approach to understanding structure-function relationships", International Joint Conference on Neural Networks, IJCNN 2009, pp. 2603–2610, © 2009 IEEE

We generalized the original network dynamics in [15] by allowing the feedback connections to affect the phase *and* amplitude of target units. We note that our objective function does not explicitly provide guidance on the projection radii for different classes of connections. The best projection radii were empirically determined. A wiring length minimization framework [27, 28] can be combined to create a constrained optimization approach.

Our results need to be confirmed through the use of larger-sized networks, and by using more realistic inputs, such as natural images. The model and analysis techniques presented in this chapter show a promising ability to address structure-function questions in neuroscience. Our model permits the exploration of several interesting avenues, including the replication of delayed feedback responses in primate cortex to illusory contours [29]; we have already reported initial results in [16]. Another simplification we made is in the polarity of the connections. Different types of feedback connections have been reported, including inhibitory and excitatory [2].

5 Functional Constraints on Network Topology

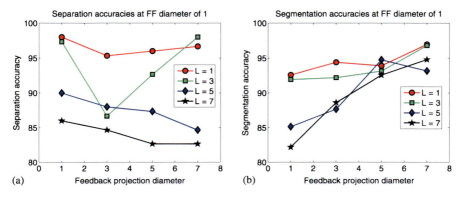

Fig. 5.11 (a) Separation accuracy plotted as a function of network connectivity using multiple 2D plots. Here, $R_{FF} = 0.5$ pixels for all cases and L is the diameter of lateral interactions. (b) Variation of segmentation accuracy when $R_{FF} = 0.5$ pixels. These plots show that separation and segmentation accuracies decrease while R_L increases

Our model does not utilize these separate types of feedback connections. Instead, we model a combined feedback effect. A more complete model would incorporate roles for these different types of feedback.

Neuroscientific measurements indicate that there are different latencies amongst connections, for example, there are longer signal delays via feedback pathways [30]. We do not explicitly model such differential signal delays. Hence, there is scope for a more sophisticated model to be built based on the one we have presented in this chapter. A more complete model will also reflect the laminar structure within each visual area such as V1 and V2, and model each layer separately [2].

5.7.1 Practical Considerations

The oscillatory network model presented in this chapter is able to perform both separation and segmentation of mixtures. The ability to perform segmentation arises from the use of oscillations, which carry a high computational cost. Indeed, other researchers have faced this bottleneck as well, which has prompted the investigation of oscillator-based hardware solutions [31]. Instead, we have chosen to implement the system on a parallel platform, an IBM p690 shared memory machine with 24 processors. The parallelization of the code was performed with the pthreads library. This allows for sufficient flexibility in performing computational experiments.

We used small images as the system input in order to demonstrate that an optimization approach is suitable for tackling the binding problem. Further research needs to be conducted in order to scale the model up to more realistic image sizes and content. We also observe that one of the requirements imposed on our model was that it should function in an unsupervised manner, due to our desire to explain biological phenomena. However, systems of practical nature are designed solve specific problems such as image retrieval, and do not carry such a biological plausibility

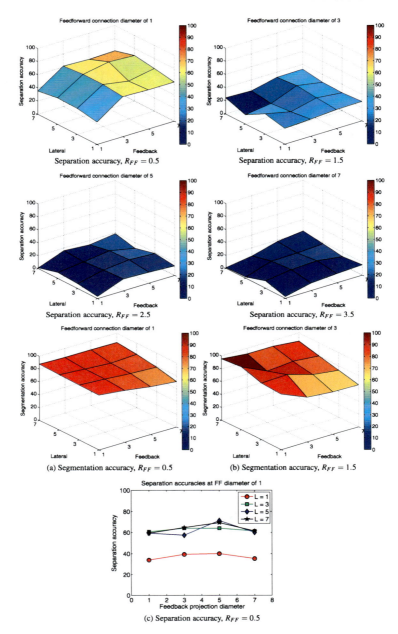

Fig. 5.12 Results obtained when uniform noise of 10% magnitude is added to the input. We do not show the segmentation accuracy for $R_{FF} = 2.5$ and $R_{FF} = 3.5$ pixels as this measure is not meaningful when the separation accuracy is very low. In (**c**), the separation accuracy for the case $R_{FF} = 0.5$ is shown via 2D plots for the sake of clarity. This plot shows that the separation accuracy for $R_L = 1.5$ is significantly higher than that for $R_L = 0.5$. Reprinted with permission, from A.R. Rao, G.A. Cecchi, "An optimization approach to understanding structure-function relationships", International Joint Conference on Neural Networks, IJCNN 2009, pp. 2603–2610, © 2009 IEEE

requirement. This implies that engineered systems can use alternate learning methods such as supervised learning to achieve superior performance. Furthermore, practical systems for image retrieval may not need to solve the binding problem because the actions they undertake may not require identification of the inputs that caused a certain classification to arise. For instance, it would be sufficient to report a single class label.

The temporal domain is essential in solving the binding problem, [32]. However, the creation of robust systems that utilize neural network based temporal approaches remains a significant challenge. This challenge needs to be addressed by the creation of the right formalization to study and model temporal neural networks. Finally, detailed experimentation is required to evaluate and improve the implementation of the models.

5.8 Conclusion

In this chapter, we presented an objective function based on sparse spatio-temporal coding in a multi-layer network of oscillatory elements. The maximization of this objective function provides a concise description of the behavior of the network units.We used this approach as a foundation to explore topological connectivity. This resulted in a network configuration that achieves separation and segmentation accuracies of 98% and 97%, respectively on a small dataset of grayscale images.

We developed quantitative metrics on the performance of the neural system, and demonstrated that structure and function questions can be answered through these metrics. We showed that the observed topological connectivity in the cortex establishes an optimum operating point for a neural system that maximizes the dual task of classification accuracy of superposed visual inputs, and their corresponding segmentation. In our interpretation, this composite task is equivalent to that required to solve the binding problem. We conclude, in consequence, that our approach provides a foundation to further explore more realistic, as well as more theoretically grounded associations between function and structure in neural systems.

Acknowledgements The authors wish to thank the management at IBM Research and the Computational Biology Center for their support of this work.

References

1. Felleman D, Essen DV (1991) Distributed hierarchical processing in the primate cerebral cortex. Cereb Cortex 1:1–47
2. Callaway E (2004) Feedforward, feedback and inhibitory connections in primate visual cortex. Neural Netw 17(5–6):625–632
3. Sporns O, Tononi G, Kotter R (2005) The human connectome: a structural description of the human brain. PLoS Comput Biol 1(4)
4. Kasthuri N, Lichtman J (2007) The rise of the 'projectome'. Nat Methods 4(4):307–308

5. Seth A, McKinstry J, Edelman G, Krichmar J (2004) Visual binding through reentrant connectivity and dynamic synchronization in a brain-based device. Cereb Cortex 14(11):1185–1199
6. Arbib MA, Erdi P (2000) Precis of neural organization: structure, function, and dynamics. Behav Brain Sci 23(4):513–533
7. Shepherd G, Stepanyants A, Bureau I, Chklovskii D, Svoboda K (2005) Geometric and functional organization of cortical circuits. Nat Neurosci 8:782–790
8. Petitot J, Lorenceau J (2003) Editorial: special issue: neurogeometry and visual perception. J Physiol 97:93–97
9. Angelucci A, Bullier J (2003) Reaching beyond the classical receptive field of v1 neurons: horizontal or feedback axons? J Physiol 97:141–154
10. Edelman S (1999) Representation and recognition in vision. MIT Press, Cambridge
11. Gray C, König P, Engel A, Singer W (1989) Oscillatory responses in cat visual cortex exhibit inter-columnar synchronization which reflects global stimulus properties. Nature 338(6213):334–337
12. Vogels TP, Rajan K, Abbott L (2005) Neural network dynamics. Annu Rev Neurosci 28:357–376
13. Burwick T (2007) Oscillatory neural networks with self-organized segmentation of overlapping patterns. Neural Computation 2093–2123
14. Von der Malsburg C (1999) The what and why of binding: the modeler's perspective. Neuron 95–104
15. Rao AR, Cecchi GA, Peck CC, Kozloski JR (2008) Unsupervised segmentation with dynamical units. IEEE Trans Neural Netw
16. Rao AR, Cecchi GA (2008) Spatio-temporal dynamics during perceptual processing in an oscillatory neural network. In: ICANN, pp 685–694
17. Rao A, Cecchi G (2010) An objective function utilizing complex sparsity for efficient segmentation in multi-layer oscillatory networks. Intl J Intell Comput Cybern
18. Peters A, Payne BR, Budd J (1994) A numerical analysis of the geniculocortical input to striate cortex in the monkey. Cereb Cortex 4:215–229
19. Budd JML (1998) Extrastriate feedback to primary visual cortex in primates: a quantitative analysis of connectivity. Proc Biol Sci 265(1400):1037–1044
20. Yoshimura Y, Sato H, Imamura K, Watanabe Y (2000) Properties of horizontal and vertical inputs to pyramidal cells in the superficial layers of the cat visual cortex. J Neurosci 20:1931–1940
21. Angelucci A, Bressloff R (2006) Contribution of feedforward, lateral and feedback connections to the classical receptive field center and extra-classical receptive field surround of primate v1 neurons. Prog Brain Res 154:93–120
22. Olshausen B, Fields D (1996) Natural image statistical and efficient coding. Network 7:333–339
23. Osipov G, Kurths J, Zhou C (2007) Synchronization in oscillatory networks
24. Przybyszewski AW (1998) Vision: does top-down processing help us to see? Curr Biol 8(4):R135–R139
25. Rao AR, Cecchi GA, Peck CC, Kozloski JR (2008) Efficient segmentation in multi-layer oscillatory networks. In: IJCNN, June, pp 2966–2973
26. Ullman S, Vidal-Naquet M, Sali E (2002) Visual features of intermediate complexity and their use in classification. Nat Neurosci 5(7):682–687
27. Durbin R, Mitchison G (1990) A dimension reduction framework for understanding cortical maps. Nature 343:644–647
28. Cherniak C, Mokhtarzada Z, Rodriguez-Esteban R, Changizi K (2004) Global optimization of cerebral cortex layout. Proc Natl Acad Sci USA 101(4):1081–1086
29. Lamme V, Supèr H, Spekreijse H (2001) Blindsight: the role of feedforward and feedback corticocortical connections. Acta Physiol 107:209–228

30. Bullier J, Hupk J, James A, Girard P (1996) Functional interactions between areas vl and v2 in the monkey. J Physiol 90(2):217–220
31. Fernandes D, Navaux P (2003) An oscillatory neural network for image segmentation. In: Progress in pattern recognition, speech and image analysis. Lecture notes in computer science, vol 2905. Springer, Berlin, pp 667–674
32. Wang D (2005) The time dimension for scene analysis. IEEE Trans Neural Netw 16(6):1401–1426

Chapter 6
Evolution of Time in Neural Networks: From the Present to the Past, and Forward to the Future

Ji Ryang Chung, Jaerock Kwon, Timothy A. Mann, and Yoonsuck Choe

Abstract What is time? Since the function of the brain is closely tied in with that of time, investigating the origin of time in the brain can help shed light on this question. In this paper, we propose to use simulated evolution of artificial neural networks to investigate the relationship between time and brain function, and the evolution of time in the brain. A large number of neural network models are based on a feedforward topology (perceptrons, backpropagation networks, radial basis functions, support vector machines, etc.), thus lacking dynamics. In such networks, the order of input presentation is meaningless (i.e., it does not affect the behavior) since the behavior is largely reactive. That is, such neural networks can only operate in the present, having no access to the past or the future. However, biological neural networks are mostly constructed with a recurrent topology, and recurrent (artificial) neural network models are able to exhibit rich temporal dynamics, thus time becomes an essential factor in their operation. In this paper, we will investigate the emergence of recollection and prediction in evolving neural networks. First, we will show how reactive, feedforward networks can evolve a memory-like function (recollection) through utilizing external markers dropped and detected in the environment. Second, we will investigate how recurrent networks with more predictable

---

This chapter is largely based on "Evolution of Recollection and Prediction in Neural Networks", by J.R. Chung, J. Kwon and Y. Choe, which appeared in the *Proceedings of the International Joint Conference on Neural Networks*, 571–577, IEEE Press, 2009.

J.R. Chung · T.A. Mann · Y. Choe (✉)
Department of Computer Science and Engineering, Texas A&M University, 3112 TAMU, College Station, TX 77843-3112, USA
e-mail: choe@cs.tamu.edu

J.R. Chung
e-mail: jchung@cse.tamu.edu

T.A. Mann
e-mail: mann@cse.tamu.edu

J. Kwon
Department of Electrical and Computer Engineering, Kettering University, 1700 W. University Avenue, Flint, MI 48504, USA
e-mail: jkwon@kettering.edu

internal state trajectory can emerge as an eventual winner in evolutionary struggle when competing networks with less predictable trajectory show the same level of behavioral performance. We expect our results to help us better understand the evolutionary origin of recollection and prediction in neuronal networks, and better appreciate the role of time in brain function.

6.1 Introduction

What is time? Since the function of the brain is closely tied in with that of time investigating the origin of time in the brain can help shed light on this question. Figure 6.1 illustrates the relationship between the past, present, and future on one hand, and brain function such as recollection and prediction on the other hand. Without recollection (or memory), the concept of past cannot exist, and likewise, without prediction, the concept of future cannot either. Furthermore, recollection seems to be a prerequisite for prediction. With this line of thought, we can reason about the possible evolutionary origin of time in the biological nervous systems. In this paper, we propose to use simulated evolution of artificial neural networks to investigate the relationship between time and brain function, and the evolution of time in the brain.

Many neural network models are based on a feedforward topology (perceptrons, backpropagation networks, radial basis functions, support vector machines, etc.), thus lacking dynamics (see [4], and selective chapters in [19]). In such networks, the order of input presentation is meaningless (i.e., it does not affect the behavior) since the behavior is largely reactive. That is, such neural networks can only operate in the present, having no access to the past or the future. However, biological neural networks are mostly constructed with a recurrent topology (e.g., the visual areas in the brain are not strictly hierarchical [15]). Furthermore, recurrent (artificial) neural network models are able to exhibit rich temporal dynamics [3, 13, 14]. Thus, time becomes an essential factor in neural network operation, whether it is natural or artificial (also see [8, 34, 36, 44]).

Our main approach is to investigate the emergence of recollection and prediction in evolving neural networks. Recollection allows an organism to connect with its past, and prediction with its future. If time was not relevant to the organism, it would always live in the eternal present.

First, we will investigate the evolution of recollection. We will see how reactive, feedforward networks can evolve a memory-like function (recollection), through utilizing external markers dropped and detected in the environment. In this part, we trained a feedforward network using neuroevolution, where the network is allowed to drop and detect markers in the external environment. Our hypothesis is that this kind of agents could have been an evolutionary bridge between purely reactive agents and fully memory-capable agents. The network is tested in a falling-ball catching task inspired by Beer [3], Ward and Ward [45], where an agent with a set of range sensors is supposed to catch multiple falling balls. The trick is that while trying to catch one ball, the other ball can go out of view of the range sensors, thus requiring some sort of memory to be successful. Our results show that even feedforward networks can exhibit memory-like behavior if they are allowed to conduct

6 Evolution of Time

Fig. 6.1 Time, recollection, and prediction. The concept of past, present, and future and brain functions such as recollection (memory) and prediction are all intricately related

some form of material interaction, thus closing the loop through the environment (cf. [37]). This experiment will allow us to understand how recollection (memory) could have evolved.

Second, we will examine the evolution of prediction. Once the recurrent topology is established, how can predictive function evolve, based on the recurrent network's recollective (memory-like) property? For this, we trained a recurrent neural network in a 2D pole-balancing task [1], again using neuroevolution (cf. [17, 26, 28]). The agent is supposed to balance an upright pole while moving in an enclosed arena. This task, due to its more dynamic nature, requires more predictive power to be successful than the simple ball-catching task. Our main question here was whether individuals with a more predictable internal state trajectory have a competitive edge over those with less predictable trajectory. We partitioned high-performing individuals into two groups (i.e., they have the same behavioral performance), those with high internal state predictability and those with low internal state predictability. It turns out that individuals with highly predictable internal state have a competitive edge over their counterpart when the environment poses a tougher problem [23].

In sum, our results suggest how recollection and prediction may have evolved, that is, how "time" evolved in the biological nervous system. We expect our results to help better understand the evolutionary origin of recollection and prediction in neuronal networks, and better appreciate the role of time in neural network models. The rest of the paper is organized as follows. Section 6.2 presents the method and results from the recollection experiment, and Sect. 6.3, those from the prediction experiment. We will discuss interesting points arising from this research (Sect. 6.4), and conclude our paper in Sect. 6.5.

6.2 Part I: Evolution of Recollection

In this section, we will investigate how memory-like behavior can evolve in a reactive, feedforward network. Below, we will describe the ball catching task, and explain in detail our neuroevolution methods for the learning component. Next, we will present the details of our experiments and the outcomes.

6.2.1 Task: Catching Falling Balls

The main task for this part was the falling ball catching task, inspired by Beer [3], Ward and Ward [45]. The task is illustrated in Fig. 6.2. See the figure caption for details. The task is simple enough, yet includes interesting dynamic components and temporal dependency. The horizontal locations of the balls are on the two different sides (left or right) of the agent's initial position. Between the left and right balls, one is randomly chosen to have faster falling speed (2 times faster than the other). The exact locations are randomly set with the constraint that they must be separated far enough to guarantee that the slower one must go out of the sensor range as the agent moves to catch the faster one. For example, as shown in Fig. 6.2C, when there are multiple balls to catch and when the balls are falling at different speeds, catching one ball (usually the faster one) results in the other ball (the slower one) going out of view of the range sensors. Note that both the left-left or right-right ball settings cannot preserve the memory requirement of the task. The vertical location, ball speed, and agent speed are experimentally chosen to guarantee that the trained agent can successfully catch both balls. In order to tackle this kind of situation, the controller agent needs some kind of memory.

The learning of connection weights of the agents is achieved by genetic search, where the fitness for an agent is set inversely proportional to the sum of horizontal separations between itself and each ball when the ball hits the ground. 10 percent of the best-performing agents in a population are selected for 1-point crossover with probability 0.9 and a mutation with the rate 0.04.

6.2.2 Methods

In order to control the ball catcher agents, we used feedforward networks equipped with external marker droppers and detectors (Fig. 6.3, we will call this the "dropper network"). The agent had five range sensors that signal the distance to the ball when the ball comes into contact within the direct line-of-sight of the sensors. We used standard feedforward networks with sigmoidal activation units as a controller (see, e.g., [19]):

$$H_j = \sigma\left(\sum_{i=1}^{N_{in}} v_{ji} I_i\right), \quad j = 1, \ldots, N_{hid},$$
$$O_k = \sigma\left(\sum_{j=1}^{N_{hid}} w_{kj} H_j\right), \quad k = 1, \ldots, N_{out},$$
(6.1)

where I_i, H_j and O_k are the activations of the i-th input, j-th hidden, and k-th output neurons; v_{ji} the input-to-hidden weights and w_{kj} the hidden-to-output weights; $\sigma(\cdot)$ the sigmoid activation function; and N_{in}, N_{hid}, and N_{out} are the number of input, hidden, and output neurons whose values are 7, 3, and 3, respectively.

The network parameters were tuned using genetic algorithms, thus the training did not involve any gradient-based adaptation. Two of the output units were used to determine the movement of the agent. If the agent was moved one step to the left when $O_1 > O_2$, one step to the right when $O_1 < O_2$, and remained in the current spot when $O_1 = O_2$.

If these were the only constructs in the controller, the controller will fail to catch multiple balls as in the case depicted in Fig. 6.2C. In order to solve this kind of

6 Evolution of Time

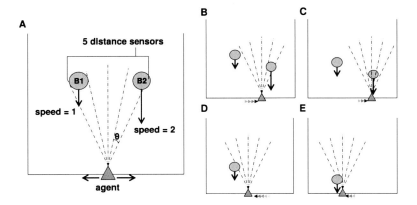

Fig. 6.2 Ball catching task. An illustration of the ball catching task is shown. The agent, equipped with a fixed number of range sensors (*radiating lines*), is allowed to move left or right at the bottom of the screen while trying to catch balls falling from the top. The goal is to catch both balls. The balls fall at different speeds, so a good strategy is to catch the fast-falling ball first (**B** and **C**) and then the go back and catch the slow one (**D** and **E**). Note that in **C** the ball on the left is outside of the range sensors' view. Thus, a memory-less agent would stop at this point and fail to catch the second ball. Adapted from [10]

problem, a fully recurrent network is needed, but from an evolutionary point of view, going from a feedforward neural circuit to a recurrent neural circuit could be nontrivial, thus our question was what could have been an easier route to memory-like behavior, without incurring much evolutionary overhead.

Our answer to this question is illustrated in Fig. 6.3. The architecture is inspired by primitive reactive animals that utilize self-generated chemical droppings (excretions, pheromones, etc.) and chemical sensors [12, 41, 46]. The idea is to maintain the reactive, feedforward network architecture, while adding a simple external mechanism that would incur only a small overhead in terms of implementation. As shown in Fig. 6.3, the feedforward network has two additional inputs for the detection of the external markers dropped in the environment, to the left or to the right (they work in a similar manner as the range sensors, signaling the distance to the markers). The network also has one additional output for making a decision whether to drop an external marker or not.

As a comparison, we also implemented a fully recurrent network, with multiple levels of delayed feedback into the hidden layer. (See [13, 14] for details.) This network was used to see how well our dropper network does in comparison to a fully memory-equipped network.

6.2.3 Experiments and Results

The network was trained using genetic algorithms (neuroevolution), where the connection weights and the dropper threshold θ were encoded in the chromosome. The

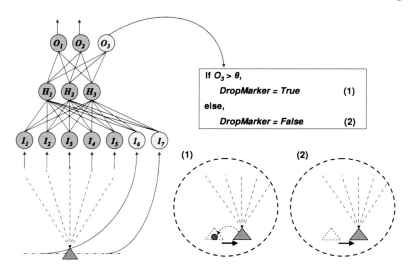

Fig. 6.3 Feedforward network with dropper/detector ("dropper network"). A feedforward network with a slight modification (dropper and detector) is shown. The basic internal architecture of the network is identical to any other feedforward network, with five range sensor (I_1 to I_5), and two output units that determine the movement (O_1 and O_2). The two added input units (I_6 and I_7) signal the presence of a dropped marker on the bottom plane, and the one additional output unit (O_3) makes the decision of whether to drop a marker at the current location or not. Note that there are no recurrent connections in the controller network itself. Adapted from [10]

fitness was inversely proportional to the sum of the distance between the agent and the ball(s) when the ball(s) contact the ground. Each individual was tested 12 times with different initial ball position (which was varied randomly) and speed (1 or 2 steps/time unit), and mixed scenarios with fast left ball vs. fast right ball. We used one-point crossover with probability 0.9, with a mutation rate of 0.04.

Figure 6.4 summarizes the main results. It is quite remarkable that feedforward networks can show an equal level of performance as that of the recurrent network, although the feedforward networks were equipped with the dropper/detector. For example, compared to the recurrent networks, the number of tunable parameters are meager for the dropper network since they do not have layers of fully connected feedback. Six additional weights for input-to-hidden, and three for hidden-to-output, plus a single threshold parameter (10 in all) is all that is needed.

One question arises from the results above. What kind of strategy is the dropper network using to achieve such a memory-like performance? We analyzed the trajectory and the dropping pattern, and found an interesting strategy that evolved. Figure 6.5 shows some example trajectories. Here, we can see a curious *overshooting* behavior.

Figure 6.6 shows how this overshooting behavior is relevant to the task, when combined with the dropping events. The strategy can be summarized as below: (1) The right ball falls fast, which is detected first. (2&3) The agent moves toward the right ball, eventually catching it (4). At this point, the left ball is outside of the

6 Evolution of Time   105

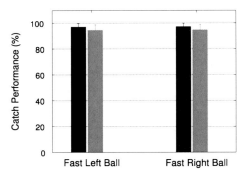

Fig. 6.4 Ball catching performance. The average ball catching performance of the dropper network is presented (*gray bar*), along with that of the recurrent network (*black bar*). The error bars indicate the standard deviation. The results are reported in two separate categories: fast left ball and fast right ball. This was to show that the network does not have any bias in performance. Both networks perform at the same high level (above 90% of all balls caught). This is quite remarkable for a feedforward network, although it had the added dropper/detector mechanism. We also tested a purely feedforward networks, but they were only able to catch 50% of the balls (catch one, miss one). Adapted from [10]

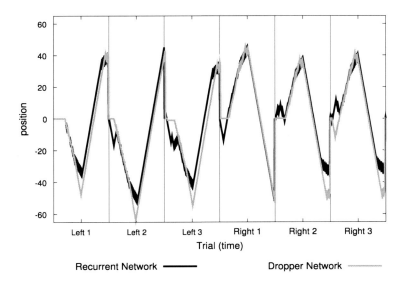

Fig. 6.5 Agent trajectory. Agent trajectory during six ball catching trials are shown (*gray*: dropper network; *black*: recurrent network). The x axis represents time, and the y axis the agent position (0 marks the initial location of the agent). Within each trial, 200 time steps are shown. As the left and the right ball positions were randomized, the peak of the trajectories differ in their y values. The first three trials were the "fast left ball" condition and the last three were the "fast right ball" condition. Both networks are successful at catching both balls within each trial, but the dropper network shows a curious *overshooting* behavior (for example, near the half way point in each trial). See Fig. 6.6 for details. Adapted from [10]

Fig. 6.6 Dropper network strategy. A strategy that evolved in the dropper network is shown. (**1**) Fast ball enters the view. (**2&3**) Agent moves toward the fast ball. (**4**) Agent catches fast ball, lose view of the slow ball, overshoots, and start dropping markers (*black dots*). (**5**) Seemingly repelled by the markers, the agent moves back to the slow ball, continuously dropping the markers, and (**6**) eventually catches it. Adapted from [10]

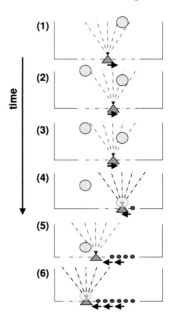

range sensors' view, it overshoots the right ball, drops a marker there, and immediately returns back, seemingly repelled by the marker that has just been dropped. (5) The agents keeps on dropping the marker while pushing back to the left, until the left ball comes within the view of the range sensor. (6) The agent successfully catches the second ball. This kind of aversive behavior is quite the opposite of what we expected, but for this given task it seem to make pretty good sense, since in some way the agent is "remembering" which direction to avoid, rather than remembering where the slow ball was (compare to the "avoiding the past" strategy proposed in [2]). Finally, we have also been able to extend our results reported here to a more complex task domain (food foraging in 2D). See [9] for details.

6.3 Part II: Evolution of Prediction

In this second part, we will now examine how predictive capabilities could have emerged through evolution. Here, we use a recurrent neural network controller in a 2D pole-balancing task. Usually recurrent neural networks are associated with some kind of memory, that is, an instrument to look back into the past. However, here we argue that it can also be seen as holding a predictive capacity, that is, looking into the future. Below, we first describe the 2D pole-balancing task, and explain our methods, followed by experiments and results. The methods and results reported in this part are largely based on our earlier work [23].

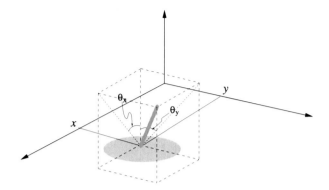

Fig. 6.7 2D pole-balancing task. The 2D pole-balancing task is illustrated. The cart (*gray disk*) with an upright pole attached to it must move around on a 2D plane while keeping the pole balanced upright. The cart controller receives the location (x, y) of the cart, the pole angle (θ_x, θ_y), and their respective velocities as the input, and generates the force in the x and the y direction. Adapted from [10]

6.3.1 Task: 2D Pole Balancing

Figure 6.7 illustrates the standard 2D pole balancing task. The cart with a pole on top of it is supposed to be moved around while the pole is balanced upright. The whole event occurs within a limited 2D bound. A successful controller for the cart can balance the pole without making it fall, and without going out of the fixed bound. Thus, the pole angle, cart position, and their respective velocities become an important information in determining the cart's motion in the immediate next time step.

6.3.2 Methods

For this part, we evolved recurrent neural network controllers, as shown in Fig. 6.8A. The activation equation is the same as Eq. 6.1, and again, we used the same neuroevolution approach to tune the weights and other parameters in the model. One difference in this model was the inclusion of a facilitating dynamics in the neuronal activation level of the hidden units. Instead of using the H_j value directly, we used the facilitated value

$$A_j(t) = H_j(t) + r\big(H_j(t) - A_j(t-1)\big), \tag{6.2}$$

where $H_j(t)$ is the hidden unit j's activation value at time t, $A_j(t)$ the facilitated hidden unit j's activation value, and r an evolvable facilitation rate parameter (see [22] for details). This formulation turned out to have a smoother characteristic, compared to our earlier facilitation dynamics in [25, 27, 29].

One key step in this part is to *measure* the predictability in the internal state dynamics. That is, given m past values of a hidden unit H_j (i.e., $\langle H_j(t-1), H_j(t-$

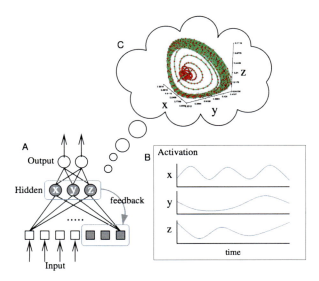

Fig. 6.8 Cart controller and its internal state. A sketch of the cart controller network is shown. (**A**) The network had 3 hidden units, which was fed back as the context input with a 1-step delay, to implement a recurrent architecture. The network had 8 inputs, each corresponding to the measures listed in Fig. 6.7. The two output units represents the force to be applied in the *x* and the *y* direction, respectively. (**B**) The activity level of the hidden units can be seen as the agent's internal state, which in this case, can be plotted as a trajectory in 3D (see **C**). Adapted from [10]

2), ..., $H_j(t - m)\rangle)$, how well can we predict $H_j(t)$. The reason for measuring this is to categorize individuals (evolved controller networks) that have a predictive potential and those that do not, and observe how they evolve. Our expectation is that individuals with more predictable internal state trajectory will have an evolutionary edge, thus opening the road for predictive functions to emerge. In order to have an objective measure, we trained a standard backpropagation network, with the past input vector $\langle H_j(t - 1), H_j(t - 2), \ldots, H_j(t - m)\rangle$ as the input and the current activation value $H_j(t)$ as the target value. Figure 6.9 shows a sketch of this approach. With this, internal state trajectories that are smoother and easier to predict (Fig. 6.10A) will be easier to train, that is, faster and more accurate, than those that are harder to predict (Fig. 6.10B). Note that the measured predictability is *not used as a fitness measure*. Predictability is only used as a post-hoc analysis. Again, the reason for measuring the predictability is to see how predictive capability can spontaneously emerge throughout evolution.

6.3.3 Experiments and Results

Figure 6.11 shows an overview of our experiment.

The pole balancing problem was set up within a 3 m × 3 m arena, and the output of the controller exerted force ranging from −10 N to 10 N. The pole was 0.5 m long,

6 Evolution of Time

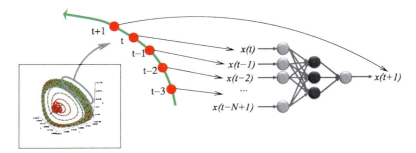

Fig. 6.9 Measuring predictability in the internal state trajectory. A simple backpropagation network was used to measure the predictability of the internal state trajectory. A sliding window on the trajectory generated a series of input vectors (N past data points) and the target values (the current data point) to construct the training set. Those with a smoother trajectory would be easier to train, with higher accuracy. Adapted from [10]

and the initial tilt of the pole was set randomly within $0.57°$. We used neuroevolution (cf. [17]). Fitness was determined by the number of time steps the controller was able to balance the pole within $\pm 15°$ from the vertical. Crossover was done with probability 0.7 and mutation added perturbation with a rate of ± 0.3. The force was applied at a 10 ms interval. The agent was deemed successful if it was able to balance the pole for 5,000 steps.

For the backpropagation predictors, we took internal state trajectories from successful controllers, and generated a training set for supervised learning, using 3,000 data points in the trajectory data. We generated an additional 1,000 inputs for validation. Standard backpropagation was used, with a learning rate of 0.2. For each data

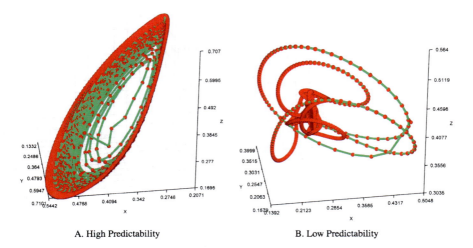

Fig. 6.10 Internal state trajectories. Typical internal state trajectories from the hidden units of the controller networks are shown for (**A**) the high predictability group and (**B**) the low predictability group

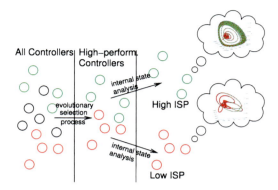

Fig. 6.11 Overview of the experiment. An overview of the experiment is shown. First, high-performing individuals (capable of balancing the pole for over 5,000 steps) are collected throughout the generations. Next, the internal state predictability of the selected ones are measured to separate the group into high internal state predictability (High ISP) and low ISP groups. The High and Low ISP groups are subsequently tested in a tougher task. Adapted from [10]

point, if the error was within 10% of the actual value, we counted that as correct, and otherwise incorrect. With this, for each trajectory we were able to calculate the predictive accuracy.

We evolved a total of 130 successful individuals, and measured their internal state predictability. Figure 6.12 shows the predictability in the 130 top individuals, which exhibits a smooth gradient. Among these, we selected the top 10 and the bottom 10, and further compared their performance. Note that since all 130 had excellent performance, the 20 that are selected in this way by definition have the same level of performance. The trick here is to put those 20 controllers in a harsher environment, by making the pole balancing task harder. We increased the initial pole angle slightly to achieve this. The results are shown in Fig. 6.13. The results show that the high internal state predictability (high ISP) group outperforms the low internal

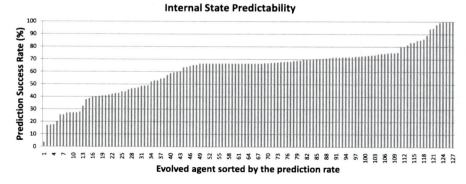

Fig. 6.12 Internal state predictability. Internal state predictability of 130 successful controllers are shown, sorted in increasing order. Adapted from our earlier work [10, 23]

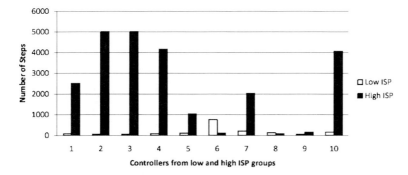

Fig. 6.13 Pole balancing performance. The performance (number of pole balancing steps) of the controller network is shown for the high ISP group (*black bars*) and the low ISP group (*white bars*). For this task the initial pole angle was increased to within $(\theta_x, \theta_y) = (0.14°, 0.08°)$. In all cases, the high ISP group does better, in many cases reaching the 5,000 performance mark, while those in the low ISP group show near zero performance. Note that these are new results, albeit being similar to our earlier results reported in [23]. Adapted from [10]

state predictability (low ISP) group by a large margin. This is a surprising outcome, considering that the two types of networks (high ISP vs. low ISP) had the same level of performance in the task they were initially evolved in. This suggests that certain internal properties, although only internally scrutinizable at one time, can come out as an advantage as the environment changes. One interesting observation we made in our earlier paper [23] is that the high performance in the high ISP group is not due to the simpler, smoother internal state trajectory linearly carrying over into simpler, smoother behavior, thus giving it an edge in pole balancing. On the contrary, we found that in many cases, high ISP individuals had complex behavioral trajectories and vice versa (see [23] for details). In sum, these results show how *predictive* capabilities could have evolved in evolving neural networks.

6.4 Discussion

The main contribution of this paper is as follows. We showed how recollection and prediction can evolve in neural circuits, thus linking the organism to its past and its future.

Our results in Part I suggest an interesting linkage between external memory and internalized memory (cf. [11, 42]). For example, humans and many other animals use external objects or certain substances excreted into the environment as a means for spatial memory (see [6, 7, 37] for theoretical insights on the benefit of the use of inert matter for cognition). In this case, olfaction (or other forms of chemical sense) serves an important role as the "detector". (Olfaction is one of the oldest sensory modalities, shared by most living organisms [20, 32, 43].) This form of spatial memory resides in the environment, thus it can be seen as external memory. On the other hand, in higher animals, spatial memory is also internalized, for example in the hippocampus. Interestingly there are several different clues that sug-

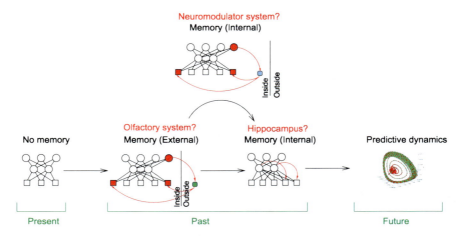

Fig. 6.14 From the present to the past, and forward to the future. Initially, only reactive behavior mediated by feedforward networks may have existed (*left-most panel, bottom*). By evolving external dropper/detector capability while maintaining the feedforward topology, simple memory function may have emerged (*second panel, bottom*), reminiscent of olfaction. Then, this kind of dropper/detector mechanism could have been internalized, resulting in something in between the dropper/detector (*second panel, bottom*) and the fully recurrent (*third panel, bottom*). Neuromodulators could be thought of as such an intermediate stage (*top panel*). Finally, a full-blown recurrent architecture may have resulted (*third panel, bottom*). The interesting thing is that the brain seem to have kept all the legacy memory systems (olfaction and neuromodulators), integrating them with the latest development (recurrent wiring)

gest an intimate relationship between the olfactory system and the hippocampus. They are located nearby in the brain, and genetically they seem to be closely related ([31, 35] showed that the Sonic Hedgehog gene controls the development of both the hippocampus and the olfactory bulb). Furthermore, neurogenesis is most often observed in the hippocampus and in the olfactory bulb, alluding to a close functional demand [16]. Finally, it is interesting to think of neuromodulators [21] as a form of internal marker dropping, in the fashion explored in this paper. Figure 6.14 summarizes the discussion above. From the left to the right, progressive augmentation to the simplest feedforward neural network that enable memory of the past and facilities for prediction of future events is shown.

Prediction (or anticipation) is receiving much attention lately, being perceived as a primary function of the brain [18, 30] (also see [38] for an earlier discussion on anticipation). Part II of this chapter raises interesting points of discussion regarding the origin and role of prediction in brain function. One interesting perspective we bring into this rich ongoing discussion about prediction is the possible evolutionary origin of prediction. If there are agents that show the same level of behavioral performance but have different internal properties, why would evolution favor one over the other? That is, certain properties internal to the brain (like high ISP or low ISP) may not be visible to the external processes that drive evolution, and thus may not persist (cf. "philosophical zombies" [5]). However, our results show that certain properties can be latent, only to be discovered later on when the changing environment helps

6 Evolution of Time   113

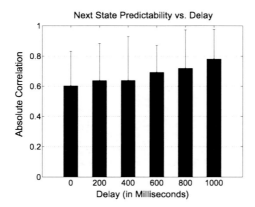

Fig. 6.15 Effect of delay on state predictability. The absolute correlations between hidden state activity at a given moment and true sensory state in the future time step are shown, for neural network controllers trained with different amount of delay in the sensory input lines (error bars indicate standard deviation). This correlation measures the degree of predictability in the internal dynamics. As the delay grows, the predictability also grows. Adapted from [33] (abstract)

bring out the fitness value of those properties. Among these properties we found prediction. Our preliminary results also indicate an important link between delay in the nervous system and the emergence of predictive capabilities: Fig. 6.15 shows that neural networks controllers exhibit more predictable behavior as the delay in the input is increased in a 2D pole-balancing task. For more discussion on the relationship between delay, extrapolation, delay compensation, and prediction; and possible neural correlates, see [24, 25, 27–29].

There are several promising future directions. For Part I, recollection, it would be interesting to extend the task domain. One idea is to allow the agent to move in a 2D map, rather than on a straight line. We expect results comparable to those reported here, and also to those in [2]. Furthermore, actually modeling how the external memory became internalized (see Fig. 6.14, second to third panel) would be an intriguing topic (a hint from the neuromodulation research such as [21] could provide the necessary insights). Insights gained from evolving an arbitrary neural network topology may also be helpful [39, 40]. As for Part II, prediction, it would be helpful if a separate subnetwork can actually be made to evolve to predict the internal state trajectory (as some kind of a monitoring process) and explicitly utilize the information.

6.5 Conclusion

In this chapter, we have shown how recollection and prediction could have evolved in neural network controllers embedded in a dynamic environment. Our main results are that recollection could have evolved when primitive feedforward nervous systems were allowed to drop and detect external markers (such as chemicals), and that prediction could have evolved naturally as the environment changed and thus conferred a competitive edge to those better able to predict. We expect our results to provide unique insights into the emergence of time in neural networks and in the brain: recollection and prediction, past and future.

References

1. Anderson CW (1989) Learning to control an inverted pendulum using neural networks. IEEE Control Syst Mag 9:31–37
2. Balch T (1993) Avoiding the past: a simple but effective strategy for reactive navigation. In: Proceedings of the 1993 IEEE international conference on robotics and automation. IEEE, New York, pp 678–685
3. Beer RD (2000) Dynamical approaches to cognitive science. Trends Cogn Sci 4:91–99
4. Bishop CM (1995) Neural networks for pattern recognition. Oxford University Press, Oxford
5. Chalmers DJ (1996) The conscious mind: in search of a fundamental theory. Oxford University Press, New York
6. Chandrasekharan S, Stewart T (2004) Reactive agents learn to add epistemic structures to the world. In: Forbus KD, Gentner D, Regier T (eds) CogSci2004. Lawrence Erlbaum, Hillsdale
7. Chandrasekaran S, Stewart TC (2007) The origin of epistemic structures and proto-representations. Adapt Behav 15:329–353
8. Choe Y, Miikkulainen R (2004) Contour integration and segmentation in a self-organizing map of spiking neurons. Biol Cybern 90:75–88
9. Chung JR, Choe Y (2009) Emergence of memory-like behavior in reactive agents using external markers. In: Proceedings of the 21st international conference on tools with artificial intelligence, 2009, ICTAI '09, pp 404–408
10. Chung JR, Kwon J, Choe Y (2009) Evolution of recollection and prediction in neural networks. In: Proceedings of the international joint conference on neural networks. IEEE Press, Piscataway, pp 571–577
11. Clark A (2008) Supersizing the mind: embodiement, action, and cognition. Oxford University Press, Oxford
12. Conover MR (2007) Predator-prey dynamics: the role of olfaction. CRC Press, Boca Raton
13. Elman JL (1990) Finding structure in time. Cogn Sci 14:179–211
14. Elman JL (1991) Distributed representations, simple recurrent networks, and grammatical structure. Mach Learn 7:195–225
15. Felleman DJ, Essen DCV (1991) Distributed hierarchical processing in primate cerebral cortex. Cereb Cortex 1:1–47
16. Frisén J, Johansson CB, Lothian C, Lendahl U (1998) Central nervous system stem cells in the embryo and adult. Cell Mol Life Sci 54:935–945
17. Gomez F, Miikkulainen R (1998) 2-D pole-balancing with recurrent evolutionary networks. In: Proceedings of the international conference on artificial neural networks. Springer, Berlin, pp 425–430
18. Hawkins J, Blakeslee S (2004) On intelligence, 1st edn. Henry Holt and Company, New York
19. Haykin S (1999) Neural networks: a comprehensive foundation, 2nd edn. Prentice-Hall, Upper Saddle River
20. Hildebrand JG (1995) Analysis of chemical signals by nervous systems. Proc Natl Acad Sci USA 92:67–74
21. Krichmar JL (2008) The neuromodulatory system: a framework for survival and adaptive behavior in a challenging world. Adapt Behav 16:385–399
22. Kwon J, Choe Y (2007) Enhanced facilitatory neuronal dynamics for delay compensation. In: Proceedings of the international joint conference on neural networks. IEEE Press, Piscataway, pp 2040–2045
23. Kwon J, Choe Y (2008) Internal state predictability as an evolutionary precursor of self-awareness and agency. In: Proceedings of the seventh international conference on development and learning. IEEE, New York, pp 109–114

24. Kwon J, Choe Y (2009) Facilitating neural dynamics for delay compensation: a road to predictive neural dynamics? Neural Netw 22:267–276
25. Lim H, Choe Y (2005) Facilitatory neural activity compensating for neural delays as a potential cause of the flash-lag effect. In: Proceedings of the international joint conference on neural networks. IEEE Press, Piscataway, pp 268–273
26. Lim H, Choe Y (2006) Compensating for neural transmission delay using extrapolatory neural activation in evolutionary neural networks. Neural Inf Process, Lett Rev 10:147–161
27. Lim H, Choe Y (2006) Delay compensation through facilitating synapses and STDP: a neural basis for orientation flash-lag effect. In: Proceedings of the international joint conference on neural networks. IEEE Press, Piscataway, pp 8385–8392
28. Lim H, Choe Y (2006) Facilitating neural dynamics for delay compensation and prediction in evolutionary neural networks. In: Keijzer M (ed) Proceedings of the 8th annual conference on genetic and evolutionary computation, GECCO-2006, pp 167–174
29. Lim H, Choe Y (2008) Delay compensation through facilitating synapses and its relation to the flash-lag effect. IEEE Trans Neural Netw 19:1678–1688
30. Llinás RR (2001) I of the vortex. MIT Press, Cambridge
31. Machold R, Hayashi S, Rutlin M, Muzumdar MD, Nery S, Corbin JG, Gritli-Linde A, Dellovade T, Porter JA, Rubin SL, Dudek H, McMahon AP, Fishell G (2003) Sonic hedgehog is required for progenitor cell maintenance in telencephalic stem cell niches. Neuron 39:937–950
32. Mackie GO (2003) Central circuitry in the jellyfish aglantha digitale iv. pathways coordinating feeding behaviour. J Exp Biol 206:2487–2505
33. Mann TA, Choe Y (2010) Neural conduction delay forces the emergence of predictive function in an evolving simulation. BMC Neurosci 11(Suppl 1):P62. Nineteenth annual computational neuroscience meeting: CNS*2010
34. Miikkulainen R, Bednar JA, Choe Y, Sirosh J (2005) Computational maps in the visual cortex. Springer, Berlin. URL: http://www.computationalmaps.org
35. Palma V, Lim DA, Dahmane N, Sánchez P, Brionne TC, Herzberg CD, Gitton Y, Carleton A, Álvarez Buylla A, Altaba AR (2004) Sonic hedgehog controls stem cell behavior in the postnatal and adult brain. Development 132:335–344
36. Peck C, Kozloski J, Cecchi G, Hill S, Schürmann F, Markram H, Rao R (2008) Network-related challenges and insights from neuroscience. Lect Notes Comput Sci 5151:67–78
37. Rocha LM (1996) Eigenbehavior and symbols. Syst Res 13:371–384
38. Rosen R (1985) Anticipatory systems: philosophical, mathematical and methodological foundations. Pergamon, New York
39. Stanley KO, Miikkulainen R (2002) Efficient evolution of neural network topologies. In: Proceedings of the 2002 congress on evolutionary computation (CEC'02). IEEE, Piscataway
40. Stanley KO, Miikkulainen R (2002) Evolving neural networks through augmenting topologies. Evol Comput 10:99–127
41. Tillman JA, Seybold SJ, Jurenka RA, Blomquist GJ (1999) Insect pheromones—an overview of biosynthesis and endocrine regulation. Insect Biochem Mol Biol 29:481–514
42. Turvey MT, Shaw R (1979) The primacy of perceiving: an ecological reformulation of perception for understanding memory. In: Nilsson L-G (ed) Perspectives on memory research: essays in honor of Uppsala University's 500th anniversary. Chap. 9. Lawrence Erlbaum Associates, Hillsdale, pp 167–222
43. Vanderhaeghen P, Schurmans S, Vassart G, Parmentier M (1997) Specific repertoire of olfactory receptor genes in the male germ cells of several mammalian species. Genomics 39:239–246
44. von der Malsburg C, Buhmann J (1992) Sensory segmentation with coupled neural oscillators. Biol Cybern 67:233–242
45. Ward R, Ward R (2006) 2006 special issue: cognitive conflict without explicit conflict monitoring in a dynamical agent. Neural Netw 19(9):1430–1436
46. Wood DL (1982) The role of pheromones, kairomones, and allomones in the host selection and colonization behavior of bark beetles. Annu Rev Entomol 27:411–446

Chapter 7
Synchronization of Coupled Pulse-Type Hardware Neuron Models for CPG Model

Ken Saito, Akihiro Matsuda, Katsutoshi Saeki, Fumio Uchikoba, and Yoshifumi Sekine

Abstract It is well known that locomotion rhythms of living organisms are generated by CPG (Central Pattern Generator). In this chapter, we discuss the synchronization phenomena and oscillatory patterns of the coupled neural oscillators using pulse-type hardware neuron models (P-HNMs) for the purpose of constructing the CPG model. It is shown that the plural coupled P-HNMs connected by excitatory-inhibitory mutual coupling can generate various oscillatory patterns. Therefore, we construct the CPG model by using the coupled P-HNMs to generate several locomotion rhythms. As a result, we show clearly that the IC chip of CPG model, which can generate the quadruped locomotion patterns, can be constructed by CMOS process. Furthermore, we implement the CPG model to the MEMS (Micro Electro Mechanical Systems) type robot for the purpose of generating locomotion.

7.1 Introduction

Oscillatory patterns of electrical activity are a ubiquitous feature in nervous systems. Living organisms use several oscillatory patterns to operate movement, swallowing, heart rhythms, etc. [1, 2]. To clarify oscillatory patterns, coupled neural oscillators are drawing attention. Synchronization phenomena or bifurcation phenomena of coupled neural oscillators have been studied using the Hodgkin–Huxley model (H–H model) or Bonhoeffer-van der Pol model (BVP model), which is a Class II neuron. Therefore, the synchronization phenomena of the coupled neural oscillators using mathematical neuron models has become the focus for the generating the oscillatory patterns of living organisms [3–5]. However, using the mathematical neuron models in large scale neural network (NN) is difficult to process in continuous time because the computer simulation is limited by the computer's performance, such as the processing speed and memory capacity. In contrast, using the hardware neuron model is advantageous because even if a circuit scale becomes large, the nonlinear operation can perform at high speed and process in continuous time. Therefore, the

K. Saito (✉) · A. Matsuda · K. Saeki · F. Uchikoba · Y. Sekine
College of Science and Technology, Nihon University, 7-24-1 Narashinodai, Funabashi-shi, Chiba, 274-8501 Japan
e-mail: kensaito@eme.cst.nihon-u.ac.jp

construction of a hardware model that can generate oscillatory patterns is desired. The hardware ring coupled oscillators have already been studied as a system which can demonstrate various oscillatory patterns and synchronization phenomena [6, 7]. For this reason, the ring coupled oscillators is expected to be a structural element of the cellular NN. However, most of the hardware models contain an inductor in circuit architecture [6–9]. If the system contains an inductor on the circuit system, it is difficult to implement the system on a CMOS IC chip.

We are studying about NN using pulse-type hardware neuron models (P-HNMs) [10–14]. The P-HNM has the same basic features of biological neurons such as threshold, refractory period, spatio-temporal summation characteristics, and enables the generation of a continuous action potentials. Furthermore, the P-HNM can oscillate without an inductor, making it easy to implement the system on a CMOS IC chip [10, 11].

Previously, we proposed coupled neural oscillators using P-HNMs (hereafter coupled P-HNM). It was constructed by cell body models and synaptic models which have excitatory and inhibitory synaptic connections. We described the oscillating phenomena in an autonomous system [13, 14].

In this chapter, firstly, we will show the circuit diagram of P-HNM and its basic characteristics. Secondly, we will discuss the synchronization phenomena and oscillatory patterns of the coupled P-HNMs. Thirdly, we will construct a CPG model, and finally, we will show the applications of our CPG model.

7.2 Pulse-Type Hardware Neuron Models

A P-HNM consists of the synaptic and cell body models. In this section, we will show the circuit diagrams and consider the basic characteristics of P-HNMs. The simulation results of oscillatory patterns are given by PSpice.

7.2.1 Synaptic Model

Figure 7.1 shows the circuit diagram of a synaptic model which has excitatory and inhibitory synaptic connections. The synaptic model has the spatio-temporal summation characteristics similar to those of living organisms. The spatial summation characteristics are realized by the adder. Moreover, the adder includes an inverting amplifier using an operational amplifier (Op-amp), the amplification factor of the inverting amplifier varies according to synaptic weight w. The temporal summation characteristics are realized by the Op-amp RC integrator, the time constant is τ. Furthermore, the inhibitory synaptic model is obtained by reversing the output of the excitatory synaptic model. Figure 7.1 shows two inputs but in the CPG model we use multi-inputs including excitatory inputs and inhibitory inputs.

Fig. 7.1 The circuit diagram of synaptic model

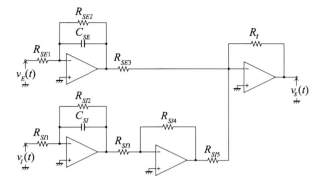

Fig. 7.2 The circuit diagram of cell body model. (Original model)

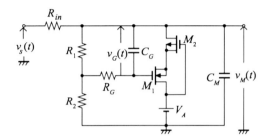

7.2.2 Cell Body Model

Figure 7.2 shows the circuit diagram of the original cell body model. The cell body model consists of a voltage control type negative resistance and an equivalent inductance, membrane capacitor C_M. The voltage control type negative resistance circuit with equivalent inductance consists of n-channel depletion-mode MOSFET, p-channel depletion-mode MOSFET, voltage source V_A, resistors R_1, R_2, and a capacitor C_G. The cell body model has the negative resistance property which changes with time like a biological neuron, and enables the generation of a continuous action potential $v_M(t)$ by a self-excited oscillation and a separately-excited oscillation. Moreover, the cell body model can switch between both oscillations by changing V_A. The separately-excited oscillation occurs by direct-current voltage stimulus or pulse train stimulus.

Figure 7.3 shows an example of the phase plane of Figure 7.2 circuit. The circuit parameters are $C_M = 90$ [pF], $C_G = 390$ [pF], $R_1 = 15$ [kΩ], $R_2 = 10$ [kΩ], $R_G = 100$ [kΩ], and $V_A = 3.5$ [V]. The abscissa is $v_M(t)$, and the ordinate is $v_G(t)$. The dotted line is $v_M(t)$-nullcline, the broken line is $v_G(t)$-nullcline, and the solid line is the attractor, as shown in Fig. 7.3. The attractor is drawn in a limit cycle. Moreover, Fig. 7.3 shows that the cell body model has the characteristic of the class II neuron such as the H–H model or BVP model. In this study, we use the cell body model as a basic element of the coupled neural oscillator.

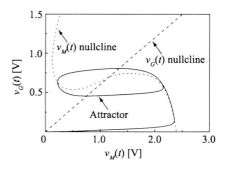

Fig. 7.3 The phase plane of P-HNM

7.2.3 CMOS Process

Figure 7.4 shows the circuit diagram of Fig. 7.2 circuit with CMOS process. We change the depletion-mode FETs of the original cell body model to enhance-mode MOSFETs. The circuit parameters are $C_G = 20$ [pF], $C_M = 10$ [pF], M_1, M_2: $W/L = 10$, M_3: $W/L = 0.1$, M_4: $W/L = 0.3$, and $V_A = 3$ [V]. However, after changing the FETs of the cell body model, the basic characteristics are equivalent to the original cell body model.

7.3 Coupled P-HNMs

7.3.1 Basic Structure of Coupled P-HNMs

Figure 7.5 shows the schematic diagram of the coupled oscillator. This figure shows that an excitatory neuron and an inhibitory neuron are coupled mutually with a synaptic model. Each neuron is self-recurrent. It is a model of the coupled neural oscillator system which enables the external inputs. In Fig. 7.5 open circles (○) indicate excitatory synaptic connections and solid circles (●) indicate inhibitory synaptic connections.

Figure 7.6 shows a circuit diagram of the coupled P-HNMs. In this section, we use the circuit of Fig. 7.4 as excitatory neuron and inhibitory neuron. The coupled

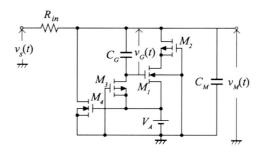

Fig. 7.4 The circuit diagram of cell body model. (CMOS process)

7 Synchronization of Coupled Pulse-Type Hardware Neuron Models

Fig. 7.5 The schematic diagram of coupled neural oscillator

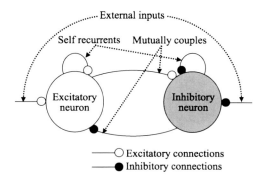

P-HNM consists of two cell body models and four synaptic models. The synaptic model's spatio-temporal dynamics summate the excitatory input and inhibitory input. The circuit parameters of the excitatory cell body model are equal to the inhibitory cell body model. In this circuit, the excitatory synaptic connection weight is as follows: $w_E = R_{Et}/R_{SE3}$, the inhibitory synaptic connection weight is as follows: $w_I = R_{It}/R_{SI5}$, the excitatory self recurrent synaptic connection weight is as follows: $w_{Er} = R_{It}/R_{Ser3}$, the inhibitory self recurrent synaptic connection weight is as follows: $w_{Ir} = R_{Et}/R_{SIr5}$. Moreover, the excitatory synaptic time constant is as follows: $\tau_E = C_{SE} R_{SE1}$, the inhibitory synaptic time constant is as follows: $\tau_I = C_{SI} R_{SI1}$, the excitatory self recurrent synaptic time constant is as follows: $\tau_{Er} = C_{SEr} R_{SEr1}$, the inhibitory self recurrent synaptic time constant is as follows: $\tau_{Ir} = C_{SIr} R_{SIr1}$. The time constant was controlled by C_{SE}, C_{SI}, C_{SEr}, and C_{SIr}.

Figure 7.7 shows an example of simulation results of phase difference characteristics between the outputs of excitatory cell body model and inhibitory cell body model. The abscissa is τ_E, and the ordinate is the phase difference θ_d. The θ_d is defined by the following equation.

$$\theta_d = \frac{t_E - t_I}{T_{EI}}, \qquad (7.1)$$

where t_E is time of peak value of the excitatory cell body model, t_I is time of peek value of the inhibitory cell body model, and T_{EI} is the pulse period of excitatory cell body model and inhibitory cell body model. We change the value of τ_I as 0.1 [μs], 1 [μs], 10 [μs] and vary the value of τ_E. The circuit parameters of cell body model are the same as Fig. 7.4. The circuit parameters of synaptic model are as follows: $R_{SE1,SE2,SE3,Et,SI1,SI2,SI3,SI4,SI5,It,SIr1,SIr2,SIr3,SIr4,SIr5,SEr1,Ser2,Ser3} = 1$ [MΩ], $R_{Ein} = R_{Iin} = 510$ [kΩ]. Synaptic weights and time constants are as follows: $w_E = w_{Er} = w_I = w_{Ir} = 1$, $\tau_E = \tau_{Er}$, $\tau_I = \tau_{Ir}$. This figure shows that the phase difference increased in the case of τ_E is increasing at $\tau_I = 0.1$ [μs]. By contrast, in the case of $\tau_I = 1, 10$ [μs] there is no phase difference.

Figure 7.8 shows an example of simulation results of phase difference characteristics by varying τ_{Ir}. The synaptic weights and time constants are as follows: $w_E = w_{Er} = w_I = w_{Ir} = 1$, $\tau_E = \tau_{Er} = 100$ [μs], $\tau_I = 0.1$ [μs]. In the case of the above parameters, the phase difference was the largest. This figure shows that the phase difference increased in the case in which τ_{Ir} is increasing.

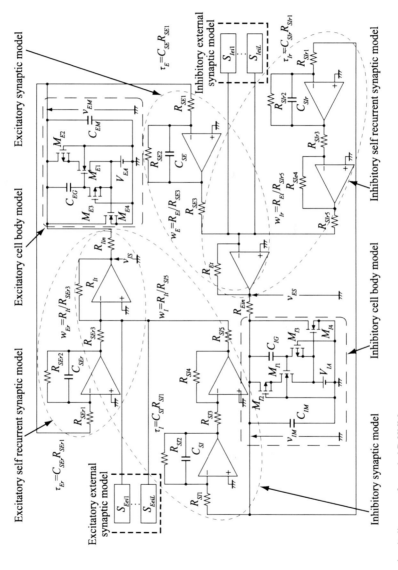

Fig. 7.6 The circuit diagram of coupled P-HNMs

Fig. 7.7 Phase difference characteristics of τ_E

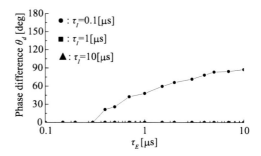

The results in Fig. 7.7, and Fig. 7.8 show that the coupled P-HNMs can adjust the phase difference between the excitatory cell body model and the inhibitory cell body model with changing synaptic time constant. Furthermore, the phase difference is larger in the case of varying τ_E than τ_{Ir}.

7.3.2 Basic Characteristics of Mutual Coupling Coupled P-HNMs

Next, we will discuss the synchronization phenomena and oscillatory patterns of mutual coupled P-HNMs by varying the synaptic time constants.

Figure 7.9 shows the schematic diagram of excitatory mutual coupling. The synaptic weights and time constants of excitatory mutual coupling are as follows: $w_{Ek1-2} = w_{Ek2-1} = 1$, $\tau_{Ek1-2} = \tau_{Ek2-1} = 0.1$ [μs]. Each synaptic weight and time constants of coupled P-HNM are as follows: $w_{I1} = w_{I2} = 2$, $w_{E1} = w_{E2} = w_{Er1} = w_{Er2} = w_{Ir1} = w_{Ir2} = 1$, $\tau_{Er1} = \tau_{Er2} = \tau_{Ir1} = \tau_{Ir2} = 1$ [μs].

Figure 7.10 shows the pattern diagram of the synchronization phenomena with excitatory mutual coupling. The abscissa is τ_{E1}, and the ordinate is τ_{E2}. In this figure, open circles (○) indicate in-phase synchronization, and solid squares (■) indicate not-in-phase synchronization. This figure shows that in this case the excitatory mutual coupling coupled P-HNM oscillates in-phase synchronization. For the other parameters, the excitatory mutual coupling coupled P-HNM oscillates not-in-phase synchronization.

Fig. 7.8 Phase difference characteristics of τ_{Ir}

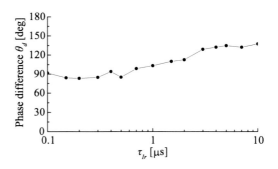

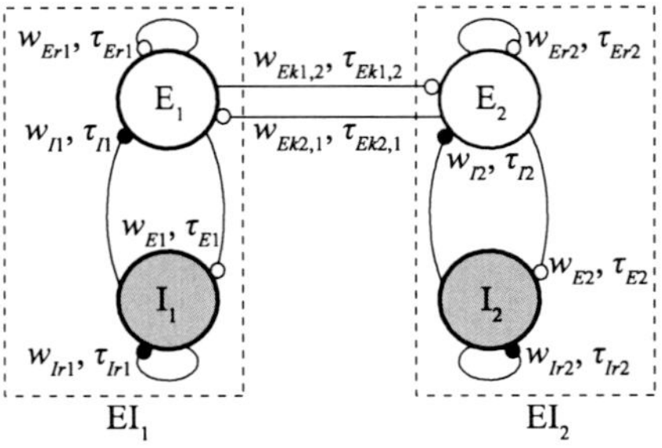

Fig. 7.9 The schematic diagram of excitatory mutual coupling

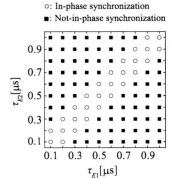

Fig. 7.10 The pattern diagram of the synchronization phenomena with excitatory mutual coupling

Figure 7.11 shows the example of the in-phase synchronization pattern of Fig. 7.10. This figure shows that the E_1 and E_2 are in-phase synchronizing in a stable state. Moreover, the not-in-phase synchronization pattern is a phase shifting pattern of Fig. 7.11.

Figure 7.12 shows the schematic diagram of inhibitory mutual coupling. The synaptic weights and time constants of inhibitory mutual coupling are as follows: $w_{Ik1-2} = w_{Ik2-1} = 1$, $\tau_{Ik1-2} = \tau_{Ik2-1} = 0.1$ [μs]. Each synaptic weight and time constants of coupled P-HNM are as follows: $w_{E1} = w_{E2} = 2$, $w_{Er1} = w_{Er2} = w_{Ir1} = w_{Ir2} = w_{I1} = w_{I2} = 1$, $\tau_{Er1} = \tau_{Er2} = \tau_{Ir1} = \tau_{Ir2} = 1$ [μs].

Figure 7.13 shows the pattern diagram of the synchronization phenomena with inhibitory mutual coupling. The abscissa is $\tau_{E1,2}$, and the ordinate is $\tau_{I1,2}$. The synaptic time constant are changed to $\tau_{E1} = \tau_{E2}$, and $\tau_{I1} = \tau_{I2}$, respectively. In this figure, open circles (○) indicate in-phase synchronization, solid circles (●) indicate

Fig. 7.11 Example of in-phase synchronization pattern of Fig. 7.10 ($\tau_{E1} = \tau_{E2} = 0.5$ [μs])

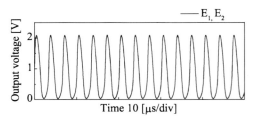

7 Synchronization of Coupled Pulse-Type Hardware Neuron Models

Fig. 7.12 The schematic diagram of inhibitory mutual coupling

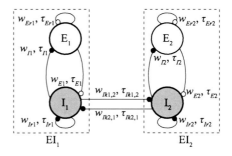

Fig. 7.13 The pattern diagram of the synchronization phenomena with inhibitory mutual coupling

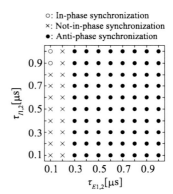

anti-phase synchronization, and crosses (×) indicate asynchronous state. In-phase synchronization occurs when the excitatory cell body model and the inhibitory cell body model have parameters of $\tau_{Er1} = \tau_{Er2} = \tau_{I1} = \tau_{I2}$. Moreover, in-phase synchronization has influence on parameters of $\tau_{E1} = \tau_{E2} = \tau_{Ik1-2} = \tau_{Ik2-1}$. Anti-phase synchronization exits when τ_{E1}, τ_{E2} are higher than 0.3 [μs]. This figure shows that anti-phase synchronization occurs because of the excitatory cell body model and the inhibitory cell body model oscillate in anti-phase synchronization. This figure shows that in-phase synchronization and anti-phase synchronization are stable at specific parameters.

Figure 7.14 shows the example of anti-phase synchronization pattern of Fig. 7.13. This figure shows that the E_1 and E_2 are anti-phase synchronizing in a stable state.

Fig. 7.14 Example of anti-phase synchronization pattern of Fig. 7.13 ($\tau_{E1,2} = \tau_{I1,2} = 0.5$ [μs])

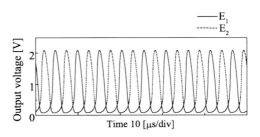

Fig. 7.15 The schematic diagram of CPG model

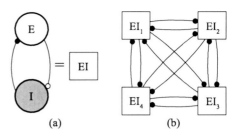

Figure 7.10, and Fig. 7.13 show that the mutual coupling coupled P-HNMs can adjust the phase difference between the coupled P-HNMs by changing the mutual coupling. Furthermore, the connections of mutual coupling can be controlled by synaptic weights. For example, synaptic weights equal to 0 means disconnection.

7.4 CPG Model

Figure 7.15 shows the schematic diagram of CPG model using coupled P-HNMs. The coupled P-HNMs are connected mutually by inhibitory synaptic model. Figure 7.15(a) shows a schematic diagram of a coupled oscillator composed of the P-HNMs. In this figure, E and I represent an excitatory neuron model and an inhibitory neuron model, respectively. The open circle indicates an excitatory connection, while the solid circle indicates an inhibitory connection. Moreover, we simplified the excitatory and inhibitory self recurrent synaptic connection. We denote the coupled oscillator by EI. Figure 7.15(b) shows a schematic diagram of the CPG model using four coupled oscillators (EI_1, EI_2, EI_3, and EI_4). Each coupled oscillator is connected to the inhibitory cell body model by the inhibitory synaptic model. Output patterns of EI_1, EI_2, EI_3 and EI_4 correspond to the left fore limb (LF), the right fore limb (RF), the right hind limb (RH), and the left hind limb (LH), respectively.

Figure 7.16 shows the relative phase difference of quadruped patterns with a typical example. In this figure, the reference of a relative phase is LF. The quadruped locomotion patterns are regarded as different modes of coordination of the limb. Therefore, it is considered that quadruped locomotion pattern transitions arise from

Fig. 7.16 Relative phase difference of quadruped patterns with a typical example

LF	RF
LH	RH

Each limb

0°	180°
270°	90°

Walk

0°	180°
0°	180°

Pace

0°	180°
180°	0°

Trot

0°	0°
180°	180°

Bound

0°	90°
270°	180°

Gallop

7 Synchronization of Coupled Pulse-Type Hardware Neuron Models

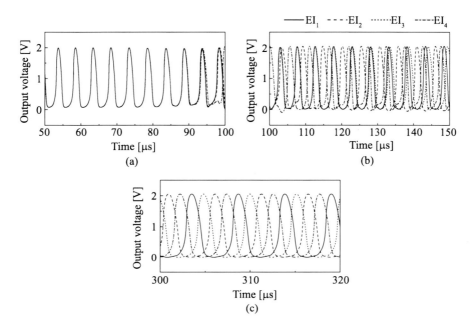

Fig. 7.17 In-phase synchronization to quadric-phase synchronization

changing cooperation of the pulse-type hardware CPG model that controls the inter limb coordination.

Figure 7.17 shows the output waveform of CPG model. The abscissa is time and the ordinate is output voltage. In this figure, the solid line indicates output waveform of EI_1, the broken line indicates output waveform of EI_2, the dotted line indicates output waveform of EI_3, and the chain line indicates output waveform of EI_4. The output waveform is the output of inhibitory cell body model of coupled P-HNM. The synaptic weights and time constants of inhibitory mutual coupling are all 1 and 1 [μs], respectively. Each synaptic weight and time constants of coupled P-HNM are all 1 and 1 [μs], respectively. The initial value of the circuit is 0, in other words the initial value of capacitance is 0. This figure shows that our CPG model outputs in-phase synchronization pattern (a), changing to asynchronous pattern (b), as a result quadric-phase synchronization (c).

Next, we will consider the method of generating quadruped patterns of our CPG model.

7.4.1 Quadruped Patterns with Varying Time Constants

Our CPG model can generate the quadruped patterns such as Fig. 7.16 with varying synaptic time constants. The advantages of varying time constants are robustness to

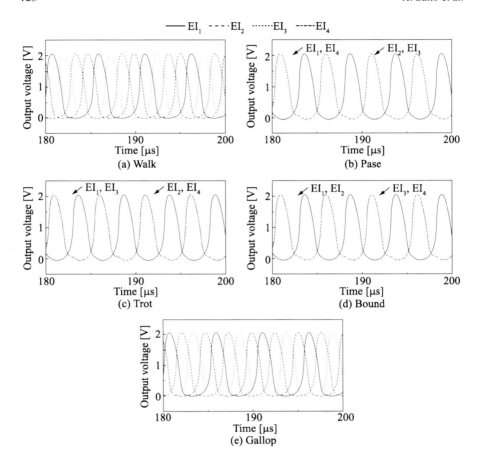

Fig. 7.18 Quadruped patterns of CPG model

noise because the coupled P-HNMs are synchronized by time constants. However, the network is stable but it is difficult to vary the capacitance in IC chip. To vary the capacitance, the external variable capacitance or variable capacitance circuit in IC chip is necessary. Moreover, the synaptic time constants of living organisms don't change.

7.4.2 Quadruped Patterns with Giving External Inputs

In this section, we will study about controlling the quadruped patterns by giving external inputs. We input external inputs to an inhibitory cell body model of coupled P-HNM. We use single pulses as external inputs which have the same relative phase difference as quadruped locomotion patterns.

7 Synchronization of Coupled Pulse-Type Hardware Neuron Models 129

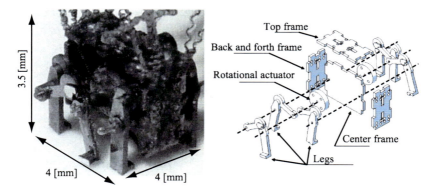

Fig. 7.19 The MEMS type robot

Figure 7.18 shows the quadruped patterns of our CPG model. The abscissa is time and the ordinate is output voltage. All the synaptic weights and time constants are the same as Fig. 7.17. This figure shows that our CPG model outputs the quadruped patterns of Fig. 7.16.

7.4.3 CPG Model for MEMS Type Robot

We construct the miniaturized robot by MEMS technology (hereafter MEMS type robot). In this section, we will show the constructed MEMS type robot. The locomotion movement of MEMS type robots can be generated by CPG model.

Figure 7.19 shows our constructed MEMS type robot. The frame components and the rotational actuators were made from silicon wafer. We use a 100, 200, 385, 500 [μm] thickness silicon wafer to construct the 4 [mm] × 4 [mm] × 3.5 [mm] size MEMS type robot. The micro fabrication of the silicon wafer was done by the MEMS technology. The shapes were machined by photolithography based ICP dry etching.

Figure 7.20 shows the schematic diagrams of the the MEMS type robot. Figure 7.20(a) shows a rotational actuator. The frame components and the rotational actuators are connected by the helical artificial muscle wire which is the shape memory alloy. Figure 7.20(b) shows a rotational actuator with 3 legs. The center leg is connected to the rotational actuator and the other legs.

Figure 7.21 shows the locomotion movement of the MEMS type robot. The MEMS type robot can move by the rotational actuator. The artificial muscle wire has a characteristic of changing length according to temperature. In the case of heating the wire shrinks and in the case of cooling the wire extends. In particular, when heating the helical artificial muscle wire rotation from A to D, the MEMS type robot will move forward.

Fig. 7.20 Schematic diagrams of the MEMS type robot

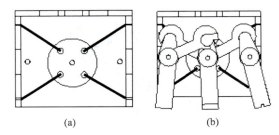

Figure 7.22 shows the schematic diagrams of the waveform to actuate the constructed MEMS type robot. To heat the helical artificial muscle, we need to input a pulse waveform such as Fig. 7.22. The pulse waveform of Fig. 7.22(a) is for forward movement, Fig. 7.22(b) is for backward movement.

Figure 7.23 shows the circuit diagram of the cell body model with driver circuit. We add the voltage follower to adjust the impedance between the cell body model and the the MEMS type robot.

Figure 7.24 shows the output waveform of the CPG model for the MEMS type robot. It is shown that our constructed CPG model can output the waveform of

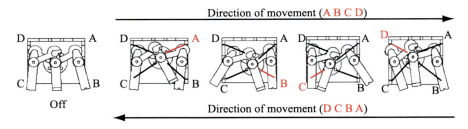

Fig. 7.21 The locomotion movement of MEMS type robot

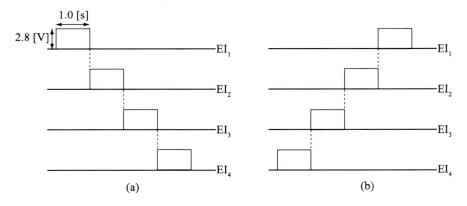

Fig. 7.22 The waveform to actuate the MEMS type robot

7 Synchronization of Coupled Pulse-Type Hardware Neuron Models

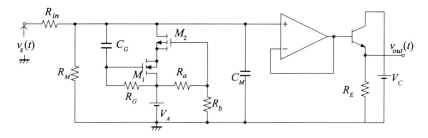

Fig. 7.23 The circuit diagram of cell body model (with driver circuit)

forward movement and backward movement such as Fig. 7.22. Thus, our coupled P-HNM is effective for CPG model for the MEMS type robot. As a result, our constructed CPG model can control the movements of the MEMS type robot.

7.4.4 CMOS IC Chip of CPG Model

Figure 7.25 shows the layout pattern of Fig. 7.15(b). This IC chip is made by ON Semiconductor Corporation CMOS 1.2 [μm] rule. The chip size is 2.3 [mm] × 2.3 [mm]. They have an 8 cell body models and 20 synaptic models in 1 chip.

Figure 7.26 shows the example of output waveform of constructed IC chip. The quadruped pattern of output waveform is bound. It is shown that our constructed IC chip can output the quadruped patterns such as Fig. 7.20. Thus, our coupled P-HNM is effective for CPG model.

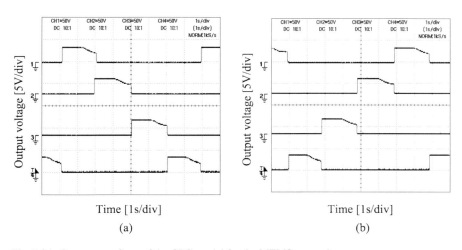

Fig. 7.24 Output waveform of the CPG model for the MEMS type robot

Fig. 7.25 The layout pattern of our CPG model

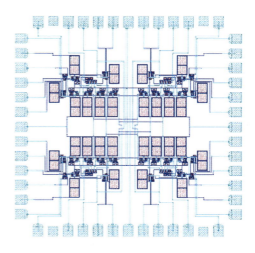

7.5 Conclusion

In this paper, we have discussed synchronization phenomena and oscillatory patterns of the coupled P-HNM for CPG Model. As a result, we developed the following conclusions.

1. P-HNM has the characteristic of class II neurons such as the H–H or BVP models. Moreover, the original model can be constructed by a CMOS process.
2. Synchronization phenomena can be controlled by changing synaptic time constant or inputting of single pulses. Furthermore, the coupled oscillator using P-HNMs can output various oscillatory patterns.
3. Constructed CPG models can output the waveform of forward movement and backward movement, which is necessary to actuate the MEMS type robot. In particular, our constructed CPG model can control the movements of the MEMS type robot.

Fig. 7.26 Output waveform of IC chip (bound)

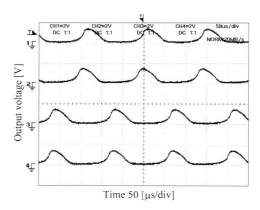

4. Constructed CMOS IC chips of CPG models can produce several quadruped pattern outputs. In addition, oscillatory patterns can be generated by giving external inputs of single pulses.

Acknowledgements The fabrication of the MEMS type robot was supported by Research Center for Micro Functional Devices, Nihon University. The VLSI chip in this study has been fabricated in the chip fabrication program of VLSI Design and Education Center (VDEC), the University of Tokyo in collaboration with On-Semiconductor, Nippon Motorola LTD, HOYA Corporation, and KYOCERA Corporation.

References

1. Delcomyn F (1980) Neural basis of rhythmic behavior in animals. Science 210:492–498
2. Arbib M (ed) (2002) The handbook of brain theory and neural networks, 2nd edn. MIT Press, Cambridge
3. Tsumoto K, Yoshinaga T, Aihara K, Kawakami H (2003) Bifurcations in synaptically coupled Hodgkin–Huxley neurons with a periodic input. Int J Bifurc Chaos 13(3):653–666
4. Tsuji S, Ueta T, Kawakami H, Aihara K (2007) Bifurcation analysis of current coupled BVP oscillators. Int J Bifurc Chaos 17(3):837–850
5. Tsumoto K, Yoshinaga T, Iida H, Kawakami H, Aihara K (2006) Bifurcations in a mathematical model for circadian oscillations of clock genes. J Theor Biol 239(1):101–122
6. Endo T, Mori S (1978) Mode analysis of a ring of a large number of mutually coupled van del Pol oscillators. IEEE Trans Circuits Syst 25(1):7–18
7. Kitajima H, Yoshinaga T, Aihara K, Kawakami H (2001) Burst firing and bifurcation in chaotic neural networks with ring structure. Int J Bifurc Chaos 11(6):1631–1643
8. Yamauchi M, Wada M, Nishino Y, Ushida A (1999) Wave propagation phenomena of phase states in oscillators coupled by inductors as a ladder. IEICE Trans Fundam, E82-A(11):2592–2598
9. Yamauchi M, Okuda M, Nishino Y, Ushida A (2003) Analysis of phase-inversion waves in coupled oscillators synchronizing at in-and-anti-phase. IEICE Trans Fundam E86-A(7):1799–1806
10. Matsuoka J, Sekine Y, Saeki K, Aihara K (2002) Analog hardware implementation of a mathematical model of an asynchronous chaotic neuron. IEICE Trans Fundam E85-A(2):216–221
11. Saeki K, Sekine Y (2003) CMOS implementation of neuron models for an artificial auditory neural network. IEICE Trans Fundam E86-A(2):248–251
12. Sasano N, Saeki K, Sekine Y (2005) Short-term memory circuit using hardware ring neural networks. Artif Life Robot 9(2):81–85
13. Nakabora Y, Saeki K, Sekine Y (2004) Synchronization of coupled oscillators using pulse-type hardware neuron models with mutual coupling. In: The 2004 international technical conference on circuits/systems, computers and communications (ITC-CSCC 2004), pp 8D2L-3-1–8D2L-3-4
14. Hata K, Saeki K, Sekine Y (2006) A pulse-type hardware CPG model for quadruped locomotion pattern. International congress series, vol 1291, pp 157–160

Chapter 8
A Universal Abstract-Time Platform for Real-Time Neural Networks

Alexander D. Rast, M. Mukaram Khan, Xin Jin, Luis A. Plana, and Steve B. Furber

Abstract High-speed asynchronous hardware makes it possible to virtualise neural networks' temporal dynamics as well as their structure. Through SpiNNaker, a dedicated neural chip multiprocessor, we introduce a real-time modelling architecture that makes the neural model run on the device independent of the hardware specifics. The central features of this modelling architecture are: native concurrency, ability to support very large ($\sim 10^9$ neurons) networks, and decoupling of the temporal and spatial characteristics of the model from those of the hardware. It circumvents a virtually fatal tradeoff in large-scale neural hardware between model support limitations or scalability limitations, without imposing a synchronous timing model. The chip itself combines an array of general-purpose processors with a configurable asynchronous interconnect and memory fabric to achieve true on- and off-chip parallelism, universal network architecture support, and programmable temporal dynamics. An HDL-like concurrent configuration software model using libraries of templates, allows the user to embed the neural model onto the hardware, mapping the virtual network structure and time dynamics into physical on-chip components and delay specifications. Initial modelling experiments demonstrate the ability of the processor to support real-time neural processing using 3 different neural models. The complete system is therefore an environment able, within a wide range of model characteristics, to model real-time dynamic neural network behaviour on dedicated hardware.

A.D. Rast (✉) · M.M. Khan · X. Jin · L.A. Plana · S.B. Furber
School of Computer Science, University of Manchester, Manchester, UK M13 9PL
e-mail: rasta@cs.man.ac.uk

M.M. Khan
e-mail: khanm@cs.man.ac.uk

X. Jin
e-mail: jinxa@cs.man.ac.uk

L.A. Plana
e-mail: plana@cs.man.ac.uk

S.B. Furber
e-mail: steve.furber@manchester.ac.uk
url: http://www.cs.manchester.ac.uk/apt

8.1 The Need for Dedicated Neural Network Hardware Support

Neural networks use an emphatically concurrent model of computation. This makes the serial uniprocessor architectures upon which many if not most neural simulators run [1] not only unable to support real-time neural modelling with large networks, but in fact architecturally unsuited to neural simulation at a fundamental level. Such concerns have become particularly pressing with the emergence of large-scale spiking models [2] attempting biologically realistic simulation of brain-scale networks. Given that biological neural networks, and likewise many interesting computational problems, demonstrate temporal dynamics, a serial computer imposing hard synchronous temporal constraints is at best a poor fit and at worst unable to model networks that change in real time. For this reason, dedicated neural network hardware embedding the concurrent model and the time dynamics into the architecture has long appeared attractive [3–5]. Yet it is also becoming clear that a fixed-model design would be a poor choice, given that just as there is debate over the architectural model in the computational community, there is no consensus on the correct model of the neuron in the biological community. The traditional digital serial model offers a critical advantage: general-purpose programmable functionality that let it simulate (if slowly) any neural network model at least in principle [6]. There has been some experimentation with hybrid approaches [7], but these impose a significant speed penalty for models and functions not integrated onto the device. Lack of flexibility is probably the main reason why neural hardware in practice has had, at best, limited success: it is not very useful to have hardware for neural modelling if the hardware forces the user to make an *a priori* decision as to the model he is going to use. If it is basic that large-scale real-time neural modelling necessitates dedicated hardware [8], it is therefore equally essential that the hardware support concurrent processing with a time model having the same level of programmability that digital computers can achieve with function. Likewise, existing software models designed for synchronous serial uniprocessor architectures are unsuitable for parallel, real-time neural hardware, and therefore a fully concurrent, hardware-oriented modelling system is a pressing need. Our proposed solution is the "neuromimetic" architecture: a system whose hardware retains enough of the native parallelism and asynchronous event-driven dynamics of "real" neural systems to be an analogue of the brain, enough general-purpose programmability to experiment with arbitrary biological and computational models. This neuromimetic system, SpiNNaker (Fig. 8.1), borrows its underlying software model and development tools from the hardware design environment, using a describe-synthesize-simulate flow to develop neural network models possessing native concurrency with accurate real-time dynamics. Our approach to practical hardware neural networks develops software and hardware with a matching architectural model: a configurable "empty stage" of generic neural components, connectivitity and programmable dynamics that take their specific form from the neural model superposed on them.

8 A Universal Abstract-Time Neural Platform

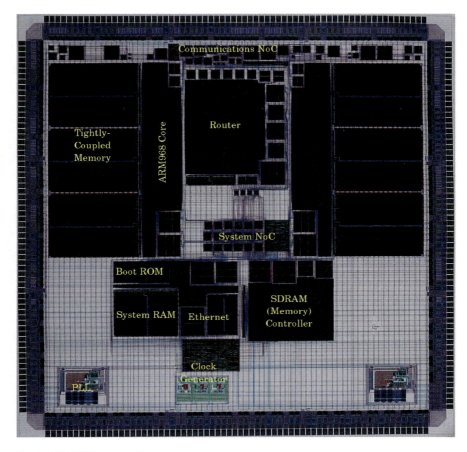

Fig. 8.1 SpiNNaker test chip

8.2 SpiNNaker: A General-Purpose Neural Chip Multiprocessor

SpiNNaker (Fig. 8.2) [9] is a dedicated chip multiprocessor (CMP) for neural simulation designed to implement the neuromimetic architecture. Its primary features are:

Native Parallelism: There are multiple processors per device, each operating completely independently from each other.

Event-Driven Processing: An external, self-contained, instantaneous signal drives state change in each process, which contains a trigger that will initiate or alter the process flow.

Incoherent Memory: Any processor may modify any memory location it can access without notifying or synchronising with other processors.

Incremental Reconfiguration: The structural configuration of the hardware can change dynamically while the system is running.

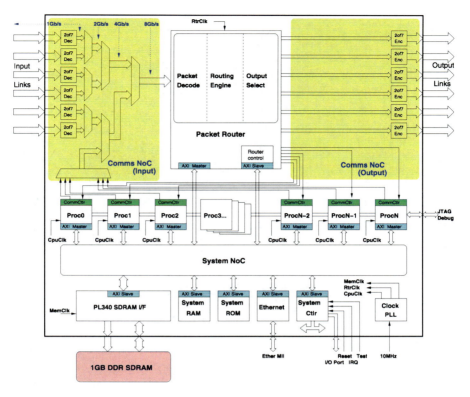

Fig. 8.2 SpiNNaker chip block diagram

A full scale SpiNNaker system envisions more than a million processing cores distributed over these CMP's to achieve a processing power of up to 262 TIPS with high concurrent inter-process communication (6 Gb/s per chip). Such a system would be able to simulate a population of more than 10^9 simple spiking neurons: the scale of a small mammalian brain. Architecturally, it is a parallel array of general-purpose microprocessors embedded in an asynchronous network-on-chip with both on-chip and inter-chip connectivity. An off-chip SDRAM device stores synaptic weights, and an on-chip router configures the network-on-chip so that signals from one processor may reach any other processor in the system, (whether on the same chip or a different chip) if the current configuration indicates a connection between the processors. The asynchronous network-on-chip allows each processor to communicate concurrently, without synchronising either to each other or to any global master clock [10]. Instead, the network uses Address-Event Representation (AER) to transmit neural signals between processors. AER is an emerging neural communication standard [11] that abstracts spikes from neurobiology into a single atomic event, transmitting only the address of the neuron that fired; SpiNNaker extends this basic standard with an optional 32-bit payload.

Since an event is an asynchronous point process with no implied temporal relationship, the chip embeds *no explicit time model*. For real-time applications, time "models itself": the clock time in the real world is the clock time in the virtual simulation, and in applications that use an abstract-time or scaled-time representation, time is superimposed onto the SpiNNaker hardware substrate through the model's configuration, rather than being determined by internal hardware components. Spatially, as well, there is no explicit topological model. A given neural connection is not uniquely identified with a given hardware link, either on-chip or between chips, so that one link can carry many signals, and the same signal can pass over many different link paths in the physical hardware topology without affecting the model connection topology of the neural network itself. Nor are processors uniquely identified with a particular neuron: the mapping is not 1-to-1 but rather many-to-one, so that in general a processor implements a collection of neurons which may be anywhere from 1 to populations of tens of thousands depending on the complexity of the model and the strictness of the real-time update constraints. SpiNNaker is therefore a completely generic device specialised for the neural application: it has an architecture which is naturally conformable to neural networks without an implementation that confines it to a specific model.

While this event-driven solution is far more universal and scalable than either synchronous or circuit-switched systems, it presents significant implementation challenges using a conventional programming methodology.

No instantaneous global state: Since communications are asynchronous the notion of global state is meaningless. It is therefore impossible to get an instantaneous "snapshot" of the system, and processors can only use local information to control process flow.

No processor can be prevented from issuing a packet: Destinations do not acknowledge source requests. Since there is no global information and no return information from destinations, no source could wait indefinitely to transmit. To prevent deadlock, therefore, processors must be able to transmit in finite time.

Limited time to process a packet at destination: Similar considerations at the destination mean that it cannot wait indefinitely to accept incoming packets. There is therefore a finite time to process any incoming packet.

Limited local memory: With 64k data memory and 32k instruction memory per processor, SpiNNaker's individual processors must operate within a limited memory space. Memory management must therefore attempt to store as much information as possible on a per-neuron rather than per-synapse basis.

Synaptic data only available on input event: Because of the limited memory, SpiNNaker stores synaptic data off-chip and brings it to the local processor only when an input event arrives. Processes that depend on the synaptic value, therefore, can only occur in the brief time after the input that the data is in local memory, and can only depend on information knowable at the time the input arrived.

No shared-resource admission control: Processors have access to shared resources but since each one is temporally independent, there can be no mechanism to prevent conflicting accesses. Therefore, the memory model is incoherent.

These behaviours, decisively different from what is typical in synchronous sequential or parallel systems, require a correspondingly different architectural model for both hardware and software. The models demonstrate much about the nature of true concurrent computation.

8.3 The SpiNNaker Hardware Model

8.3.1 SpiNNaker Local Processor Node: The Neural Module

The local processor node (Fig. 8.3) realises native parallelism. As the system building block, it is the on-chip hardware resource that implements the neural model. SpiNNaker uses general-purpose low-power ARM968 processors to model the neural dynamics. Each processor also contains a high-speed local Tightly Coupled Memory (TCM), arranged as a 32K instruction memory (ITCM) and a 64K data memory (DTCM). A single processor does not implement a single neuron but instead a group of neurons (with number dependent on the complexity of the model); running at 200 MHz a processor can simulate about 1000 simple yet biologically plausible neurons such as [12], using the ITCM to contain the code and the DCTM the neural state data. We have optimised SpiNNaker for spiking neural networks, with an execution model and process communications optimised for concurrent neural processing rather than serial-dominated general-purpose computing. In the spiking model, neurons update their state on receipt of a spike, and likewise output purely in the form of a spike (whose precise "shape" is considered immaterial), making it possible to use event-driven processing [6]. An on-board communications controller embedded with each core controls spike reception and generation for all the neurons being simulated on its associated processor. In addition, the processor contains a programmable hardware timer, providing a method to generate single or repeated absolute real-time events. The processor node does not contain the synaptic memory, which resides instead in off-chip SDRAM and is made "virtually local" through an integrated DMA controller. The local node therefore, rather than being a fixed-function, fixed-mapping implementation of a neural network component, appears as a collection of general-purpose event-driven processing resources.

8.3.2 SpiNNaker Memory System: The Synapse Channel

The second subsystem, the synapse channel, is an *incoherent* memory resource, that is, there is no mechanism to *enforce* data consistency between processor accesses. It implements the synaptic model. Since placing the large amount of memory required for synapse data on chip would consume excessive chip area, we use an off-the-shelf SDRAM device as the physical memory store and implement a linked chain of components on-chip to make synapse data appear "virtually local" by swapping

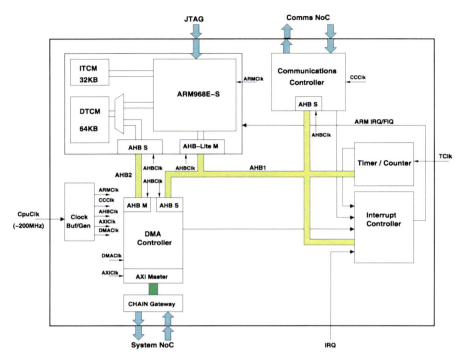

Fig. 8.3 ARM968E-S with peripherals

it between global memory and local memory within the interval between events that the data is needed. The critical components in this path are an internal asynchronous Network-on-Chip (NoC), the System NoC, connecting master devices (the processors and router) with slave memory resources at 1 GB/s bandwidth, and a local DMA controller per node able to transfer data over the interface at 1.6 GB/s in sequential burst requests. We have previously demonstrated [13] that the synapse channel can transfer required synaptic weights from global to local memory within a 1 ms event interval while supporting true concurrent memory access so that concurrent requests from different processors remain non-blocking. This makes it possible for the processor to maintain real-time update rates. Analogous to the process virtualisation the neural module achieves for neurons, the synapse channel achieves memory virtualisation by mapping synaptic data into a shared memory space, and therefore not only can SpiNNaker implement multiple heterogeneous synapse models, it can place these synapses anywhere in the system and with arbitrary associativity.

8.3.3 SpiNNaker Event-Driven Dynamics: The Spike Transaction

SpiNNaker has an event-driven processing architecture. An *event*: a point process happening in zero time, is the unit of communication that implements the tem-

Fig. 8.4 SpiNNaker AER spike packet format. Spike packets are usually type MC. Types P2P and NN are typically for system functions

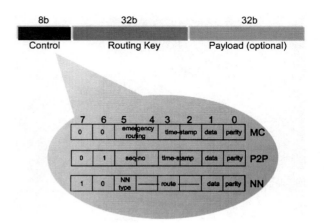

poral model. SpiNNaker uses a vectored interrupt controller (VIC) in each processor core to provide event notification to the processor. Events are processor interrupts and are of 2 principal types. The more important type—and the only one visible to the neural model—is the spike event: indication that a given neuron has signalled to some neuron the current processor is modelling. Spikes use the AER abstraction of the actual spike in a biological neuron that simplifies it to a zero-time event containing information about the source neuron, and possibly a 32-bit data payload (Fig. 8.4). The second type is the process event: indication of the completion of an internal process running on a hardware support component. Process events make possible complete time abstraction by freeing the processor from external synchronous dependencies. Within the ARM CPU, a spike event is a Fast Interrupt Request (FIQ) while a process event is an Interrupt Request (IRQ) [14]. To program a time model the user programs the VIC, interrupt service routine (ISR), and, if needed, the internal timer. Since each processor has its own independent VIC, ISR, and timer, SpiNNaker can have multiple concurrent time domains on the same chip or distributed over the system. It is because interrupts, and hence events, are asynchronous, that is, they can happen at any time, that SpiNNaker can have a fully abstract, programmable time model: interrupt timing sets the control flow and process sequencing independently of internal clocks.

8.3.4 SpiNNaker External Network: The Virtual Interconnect

The external network is a "topological wireframe" supporting incremental reconfiguration. It is therefore the hardware resource that implements the model netlist. Signals propagate over a second asynchronous NoC, the Communications NoC [10], supporting up to 6 Gb/s per chip bandwidth [9] that connects each processor core

Fig. 8.5 Multichip SpiNNaker CMP system

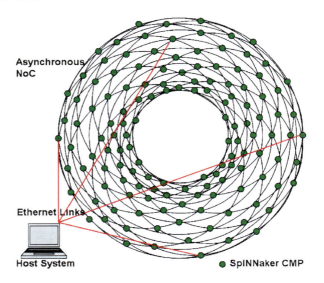

on a chip to the others, and each chip with six other chips. The hub of the NoC is a novel on-chip multicast router that routes the packets (spikes) to the internal on-chip processing cores and external chip-to-chip links using source-based associative routing. Each chip's Communications NoC interface links to form a global asynchronous packet-switching network where a chip is a node. A system of any desired scale can be formed by linking chips to each other with the help of these links, continuing this process until the system wraps itself around to form a toroidal mesh of interconnected chips as shown in Fig. 8.5.

Because the NoC is a *packet-switched* system, the physical topology of the hardware is completely independent of the connection topology of the network being modelled. To map physical links into neural connections, each chip has a configurable router containing 1024 96-bit associative routing entries that specify the routes associated with any incoming packet's routing key. A default routing protocol for unmatched inputs in combination with hierarchical address organisation minimises the number of required entries in the table. By configuring the routing tables (using a process akin to configuring an FPGA) [15], the user can implement a neural model with arbitrary network connectivity on a SpiNNaker system. Since, like the System NoC, the Communications NoC is asynchronous, there is no deterministic relation between packet transmission time at the source neuron and its arrival time at the destination(s). Spike timing in the model is a function of the programmed temporal dynamics, independent of the specific route taken through the network. Once again this decouples the communications from hardware clocks and makes it possible for a SpiNNaker system to map (virtually) any neural topology or combination of topologies to the hardware with user-programmable temporal model.

8.4 The SpiNNaker Software Model

8.4.1 3-Level System

From the point of view of the neural modeller, SpiNNaker hardware is a series of generic processing blocks capable of implementing specific components of neural functionality. The user would typically start with a neural model description which needs to be transformed into its corresponding SpiNNaker implementation. Modellers will most likely not be familiar with, or necessarily even interested in, the low-level details of native SpiNNaker object code and configuration files. Users working at different levels of abstraction therefore need an automated design environment to translate the model description into hardware object code to load to the device. We have created a software model (Fig. 8.6) based on the flow of hardware description language (HDL) tools, that use a combination of synthesis-driven instantiation [16] (automated generation of hardware-level netlists using libraries of templates that describe implementable hardware components) and concurrent simulation environments [17] (software that uses a parallel simulation engine to run multiple processes in parallel). This environment uses SystemC [18] as its base, the emerging standard for high-level hardware modelling. SystemC has several important advantages. It is a concurrent language, matching well the massive parallelism of neural networks. It contains a built-in simulator that eliminates the need to build one from the ground up. It has also been designed to abstract hardware details while providing cycle-accurate realism if necessary, making it possible to target specific hardware with a behavioural model while retaining accurate temporal relationships at each stage. Critically, SystemC supports the use of class templating, the ability to describe an object generically using the template parameters to specify the particular implementation. This makes it possible to use the same model at different levels of abstraction simply by changing the template parameter. The software environment defines 3 levels of abstraction: device level, system level, and model level. At the device level, software functions are direct device driver calls written mostly in hand-coded assembly that perform explicit hardware operations without reference to the neural model. The system level abstracts device-level functions to neural network functions, implementing these functions as SpiNNaker-specific operation sequences: templates of neural functionality that invoke a given hardware function. At the model level, there is no reference to SpiNNaker (or any hardware) as such; the modeller describes the network using abstract neural objects that describe broad classes of neural behaviour. In principle, a network described at the model level could be instantiated on any hardware or software system, provided the library objects at their corresponding system and device levels existed to "synthesize" the network into the target implementation. This 3-level, HDL-like environment allows modellers to develop at their own level of system and programming familiarity while retaining the native concurrency inherent to neural networks and preserving spatiotemporal relations in the model.

Fig. 8.6 SpiNNaker system software flow. *Arrows* indicate the direction in which data and files propagate through the system. A *solid line* represents a file, where *dashed lines* indicate data objects. *Boxes* indicate software components, the *darker boxes* being high-level environment and model definition tools, the *lighter ones* hardware-interfacing components that possess data about the physical resources on the SpiNNaker chip

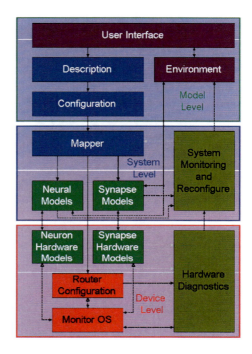

8.4.2 Model Level: Concurrent Generic Description

The model level considers a neural network as a process abstraction. The user describes the network as an interaction between 2 types of containers: neural objects and synaptic objects. Both types of containers represent groups of individual components with similar behaviour: for example, a neural object could represent a group of 100 neurons with identical basic parameters. This makes it possible to describe large neural networks whose properties might be specified statistically by grouping components generated from the same statistical distribution within a single object. Although we use the terms "neural" and "synaptic" for convenience, a neural object need not necessarily describe only neurons: the principal difference between the objects is that the synaptic object is an active SystemC *channel* and therefore defines a communication between processes whereas a neural object is a *module*. These objects take template parameters to describe their functionality: object classes that define the specific function or data container to implement. The most important of these classes are functions (representing a component of neural dynamics), signal definitions (determining the time model and data representation) and netlists (to represent the connectivity). Thus, for example, the modeller might define a set of differential equations for the dynamic functions, a spike signal type, and a connection probability to generate the netlist, and instantiate the network by creating neural and synaptic objects referencing these classes in their templates. At the model level, therefore, specifying time is entirely a matter of the template definition: the user determines the time model in the specification of the dynamic functions and the signal type.

Processes execute and communicate concurrently using SystemC's asynchronous event-driven model. For spiking neural networks, the event-driven abstraction is obvious: a spike is an event, and the dynamic equations are the response to each input spike. New input spikes trigger update of the dynamics. In nonspiking networks, different abstractions are necessary. One easy and common method is time sampling: events could happen at a fixed time interval, and this periodic event signal triggers the update. Alternatively, to reduce event rate with slowly-variable signals, a neuron may only generate an event when its output changes by some fixed amplitude. For models with no time component, the dataflow itself can act as an event: a neuron receives an input event, completes its processing with that input, and sends the output to its target neurons as an event. The important point to observe is: decisions about the event representation at the Model level could be almost entirely arbitrary. Since at this level the model is a process abstraction, it could run, in principle, on any hardware platform that supports such asynchronous communications (as SpiNNaker does) while hiding low-level timing differences between platforms.

8.4.3 System Level: Template Instantiation of Library Blocks

At the system level, the model developer gains visibility of the neural functions SpiNNaker is able to implement directly. System-level models can be optimised for the actual hardware, and therefore can potentially run faster; however, they run more slowly in software simulation because of the need to invoke hardware-emulation routines. Our approach uses template parameters to access the hardware components. A given system-level object is a generalised neural object similar to a model-level object, whose specific functionality comes from the template. At the system level, however, a template is a hardware "macro"—for example, a function GetWeights() that requests a DMA transfer, performs the requisite memory access, and retrieves a series of weights, signalling via an interrupt when complete. Time at the system level is still that of the neural model, reflecting the fact that the only hardware event visible is the spike event.

The user at system level specifies time as an input argument to the template functions: the "real time" the process would take to complete in the model. The processor can then arbitrarily reorder actual hardware timing as necessary to optimise resource use while preserving real-time-accurate behaviour. We have earlier shown [19] how to use this reordering capability to achieve accurate neural updating and STDP synaptic plasticity in an event-driven system. Both processes are interrupt-driven routines with a deferred process, programmed into instruction memory. At this level, events are transactions between objects representing individual components. Responses to events are the subroutine calls (or methods) to execute when the event arrives. These methods or functions will be different for different neural models, and because automated tools must be able to associate a given model with a given series of system objects, the System level is mostly a collection of hardware libraries for different neural models. Each library defines the event representation as set of source component functions: a Packet-Received event, a DMA event, a Timer event, and an Internal (processor) event. It must also account for important system properties: in the case of SpiNNaker, no global state information and one-way communication. System level descriptions are the source input for the SpiNNaker "synthesis" process: a bridge between the model level and the device level that uses templates as the crucial link to provide a SpiNNaker hardware abstraction layer.

8.4.4 Device Level: A Library of Optimised Assembly Routines

The device level provides direct interfacing to SpiNNaker hardware as a set of event-driven component device drivers. In the "standard SpiNNaker application model" events are interrupts to the neural process, triggering an efficient ISR to call the developer-implemented system-level neural dynamic function associated with each event. ISR routines therefore correspond directly to device-level template parameters. Time at device level is the "electronic time" of the system, as opposed to the "real time" of the model. The device level exposes the process events as well as the spike event, and therefore the programmer specifies a time model by explicit configuration of the hardware devices: the timer, the DMA controller, and the communications controller, along with the ARM968 assembly code. The hardware packet encoding is visible along with the physical registers in the DMA and communications controllers. Most of the device level code is therefore a series of interrupt-driven device drivers acting as support functions for the system level. Since device level code does not consider the neural model, these drivers are common across many models (and libraries), and includes operating-system-like system support, startup and configuration routines essential for the operation of the chip as a whole, but irrelevant from the point of view of the model.

We have implemented an initial function library as part of the configuration process for the SpiNNaker system using ARM968 assembly language for optimal performance. This device driver library includes the functions needed by a neural application where it has to interact with the hardware to model its dynamics. We have optimised the functions to support real-time applications in an event-driven model with efficient coding and register allocation schemes [20]. We have also optimised the SpiNNaker memory map so that the processor will start executing the ISR in just one cycle after receiving the event. In our reference application sending a spike requires 4 ARM instructions while receiving a spike requires 27 instructions, including the DMA request to upload the relevant data block into the local memory. By providing a ready-made library of hardware device drivers, we have given users access to carefully optimised SpiNNaker neural modelling routines while also presenting a template for low-level applications development should the user need to create his own optimised hardware drivers for high-performance modelling.

8.5 SpiNNaker System Design and Simulation

8.5.1 Design of Hardware Components

The SpiNNaker chip is a GALS system [10]—that is, a series of synchronous clocked modules embedded in an asynchronous "sea". Such a system typically requires a mix of design techniques. To minimise development time, we have attempted where possible to use industry-standard tool flows and off-the-shelf componentry. Most of the processor node, including the ARM968 and its associated interrupt controller and timers, along with the memory interface, are standard IP blocks available from ARM. A further set of blocks: the communications controller, the DMA controller, and the router, were designed in-house using synchronous design flow with HDL tools. We implemented these components using Register Transfer Level (RTL)-level descriptions in Verilog and tested them individually using the industry-standard Synopsys VCS concurrent Verilog simulation environment. Finally, the asynchronous components: the NoC's and the external communications link, used a combination of synthesis-like tools from Silistix and hand-designed optimisation to achieve required area and performance constraints. Where component design has used low-level cycle-accurate tools, system-level hardware testing and verification, by contrast, is being done using higher-level SystemC tools.

8.5.2 SystemC Modelling and Chip-Level Verification

One of the main objectives of this work has been to provide an early platform to develop and test applications for SpiNNaker while the hardware is still in the design phase. It is possible to verify the cycle accurate behaviour of individual components using HDL simulation. However, verifying the functional behaviour of a neural computing system on the scale of SpiNNaker would be unmanageably complex, either to demonstrate theoretically, or to simulate using classical hardware description languages such as VHDL and Verilog. In addition, industry-standard HDL simulators emphasize synchronous design, making verification of asynchronous circuits difficult and potentially misleading. Therefore, as part of the SpiNNaker project, we have created a SystemC system-level model for the SpiNNaker computing system to verify its functional behaviour—especially the new communications infrastructure. SystemC supports a higher level of timing abstraction: the Transaction Level Model (TLM), exhibiting "cycle approximate" behaviour. In a TLM-level simulation, data-flow timing remains accurate without requiring accuracy at the level of individual signals. This makes it possible on the one hand to integrate synchronous and asynchronous components without generating misleading simulation results, and on the other to retain timing fidelity to the neural model since the data-flow timing entirely determines its behaviour. We developed SystemC models for in-house components [21] and integrated them with a cycle-accurate instruction set simulator for the ARM968E-S processor and its associated peripherals using ARM

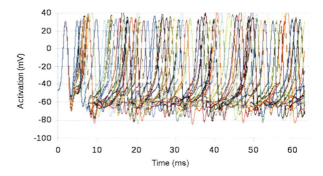

Fig. 8.7 SpiNNaker top-level model output of the spiking network. For clarity, only 25 neurons are shown

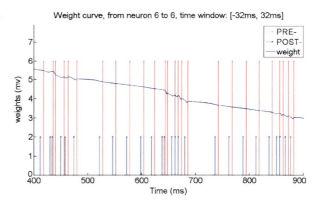

Fig. 8.8 Weight modification from the temporal correlation of the pre- and post-synaptic spike time. Pre-synaptic spikes trigger the modification

SoC Designer. SoC Designer does not support real-time delays, therefore we captured the behaviour of the asynchronous NoC in terms of processor clock cycles. We then tested the complete system behaviour extensively in two neural application case studies [9]. For simplicity of simulation on a host PC with limited memory, the model simulates 2 processing cores per chip—the number of cores on the initial test chip. With all chip components in the simulation, we were able to achieve a simulation of 9 chips running concurrently, thus verifying both on-chip and inter-chip behaviour. With system-level functional verification, we have thus been able to achieve both a strong demonstration of the viability of the SpiNNaker platform for real-world neural applications and a methodology for the development and testing of new neural models prior to their instantiation in hardware.

8.5.3 Performance and Functionality Testing

To verify the event-driven model and develop the neural library routines, we used an ARM968 emulation built with ARM's SOC designer, containing one processing node together with its associated peripherals: DMA controller, router, interrupt controller, and memory system. Router links wrap around connecting outputs to inputs,

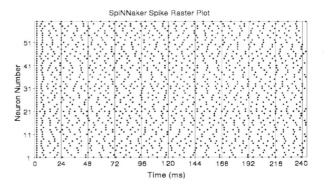

Fig. 8.9 SpiNNaker spiking neural simulation raster plot. The network simulated a random network of 60 neurons, each given an initial impulse at time 0. To verify timing, synaptic plasticity was off

so that all packets go to the emulated processor. We successfully tested a reference neural network using Izhikevich [12] neural dynamics for 1,000 neurons, with random connectivity, initial states, and parameters, updating neural state once per ms. Using assembly code programming and 16-bit fixed point arithmetic (demonstrated in [22], [23]), it takes 8 instructions to complete one update. Modelling 1,000 neurons with 100 inputs each (10% connectivity) firing at 10 Hz, requires 353 μs SpiNNaker time to simulate 1 ms of neural behaviour. Modelling 1 ms for 1,000 neurons with 1,000 inputs each (100% connectivity) firing at 1 Hz requires approximately the same computational time. The model can therefore increase the connectivity by reducing the firing rates as real neural network systems do, without losing real-time performance. This model ran on the SystemC model with a small population of neurons, reproducing the results presented in [24].

After testing neuronal dynamics, we created a second network, largely based on the network in [25] to verify synaptic dynamics using the STDP model we described in [19] and [26]. The network has 48 Regular Spiking Excitatory neurons and 12 Fast Spiking Inhibitory neurons. Each neuron connects randomly to 40 neurons (self-synapses are possible) with random 1–16 ms delay; inhibitory neurons only connect to excitatory neurons. Initial weights are 8 and −4 for excitatory and inhibitory connections respectively. The time window for STDP was $\tau_+ = \tau_- = 32$ ms, with update strength $A_+ = A_- = 0.1$. Inhibitory connections are not plastic [27]. There are 6 excitatory and 1 inhibitory input neurons, receiving constant input current $I = 20$ nA to maintain a high firing rate.

Figure 8.8 gives the detailed weight modification from a simulation with a duration of 10 sec in biological time. We selected the self-connection for neuron id 6 (w_{66}, an input neuron) to produce a clear plot. The graph shows the correlation between weight update and the pre- and post-synaptic timing.

Using the results from the neuronal dynamics experiment, we ran a cycle-accurate simulation of the top-level model (Fig. 8.7) to analyse the impact of system delays on spike-timing accuracy. Figures 8.9 and 8.10 show the results. Each point in the raster plot is one spike count in the histogram. The spike-raster test verified that there are no synchronous system side effects that systematically affect the model timing. We used the timing data from the simulation to estimate the timing error for each of the spikes in the simulation—that is, the difference between the

Fig. 8.10 SpiNNaker spike error histogram. Estimated spike-timing errors are from the same network as the raster plot

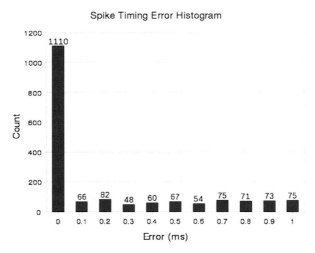

model-time "actual" timing of the spike and the system-time "electronic" timing of the spike—by locally increasing the timing resolution in the analytic (floating-point) Izhikevich model to 50 µs in the vicinity of a spike and recomputing the actual spike time. Most spikes (62%) have no timing error, and more than 75% are off by less than 0.5 ms—the minimum error necessary for a spike to occur ±1 update off its "true" timing. Maximum error is, as expected, 1 ms, since the update period fixes a hard upper bound to the error. In combination with the raster plot verifying no long-term drift, the tests indicate that SpiNNaker can maintain timing fidelity within a 1 ms resolution.

As a test of a larger, multichip simulation, we implemented a 4-chip system with 2 processors per chip. This system corresponding to the hardware configuration of the first test board. To test the system in a "real-world" application, we created a synthetic environment: a "doughnut hunter" application. The network in this case had visual input and motion output; the goal was to get the position of the network's (virtual) body to a target: an annulus or "doughnut". Testing (Figs. 8.11, 8.12, and 8.13) verified that the network could successfully track and then move its body towards the doughnut, ultimately reaching the target. These test verify the overall system functionality: the neural model behaved as expected both at the behavioural level and at the signal (spike) level.

8.6 Neural Modelling Implications

An important observation of the tests is that in a system with asynchronous components, the behaviour is nondeterministic. While most of the spikes occurred without timing error, some had sufficient error to occur at the next update interval. The effect of this is to create a ±1 ms timing jitter during simulation. Biological neural networks also exhibit some jitter, and there is evidence to suggest that this random

Fig. 8.11 Far away from the target

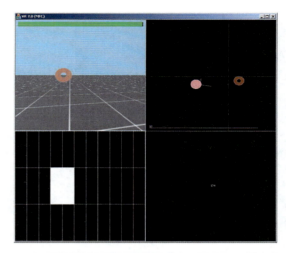

Fig. 8.12 Approaching the target

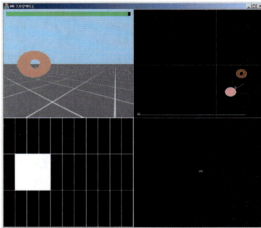

Fig. 8.13 Doughnut hunter test. Successive frames show the network's "body" as it approaches the target. Above is when the network reaches the target

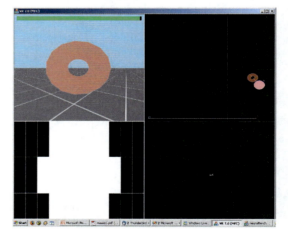

8 A Universal Abstract-Time Neural Platform

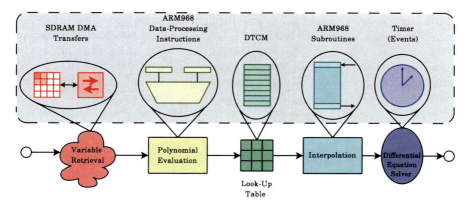

Fig. 8.14 A general event-driven function pipeline for neural networks. The *grey box* is the SpiNNaker realisation

phase error may be computationally significant [28]. It is not clear whether asynchronous communications replicates the phase noise statistics of biological networks, but its inherent property of adding some phase noise may make it a more useful platform for exploration of these effects than purely deterministic systems to which it is necessary to add an artificial noise source. In addition, the modeller can (statically) tune the amount of noise, to some degree, by programming the value of the update interval acting as an upper bound on the phase noise. A GALS system like SpiNNaker therefore appears potentially capable of reproducing a wider range of neural behaviours than traditional synchronous systems, while supporting the programmability that has been a limiting factor in analogue neuromorphic devices.

It has now been possible to test 3 different models: a spiking system with Izhikevich neurons and synapses having spike-timing-dependent plasticity (STDP), a classical multilayer perceptron (MLP) with delta-rule synapses using backpropagation, and a leaky-integrate-and-fire spiking neuron using fixed connections. From the models that have successfully run it is clear that SpiNNaker can support multiple, very different neural networks; how general this capability is remains an important question. We can define a generalised function pipeline that is adequate for most neural models in existence (Fig. 8.14). The pipeline model emerges from a consideration of what hardware can usually implement efficiently in combination with observations about the nature of neural models. Broadly, most neural models, at the level of the atomic processing operation, fall into 2 major classes, "sum-and-threshold" types, that accumulate contributions from parallel inputs and pass the result through a nonlinearity, and "dynamic" types, that use differential state equations to update internal variables. The former have the general form $S_j = T(\sum_i w_{ij} S_i)$ where S_j is the output of the individual process, T is some nonlinear function, i are the input indices, w_{ij} the scaling factors (usually, synaptic weights) for each input, and S_i the inputs. The latter are systems with the general form $\frac{dX}{dt} = E(X) + F(Y) + G(P)$ where E, F, and G are arbitrary functions, X is a given process variable, Y the other variables, and P various (constant) parameters. Meanwhile, SpiNNaker's processors can easily implement polynomial functions but other types, for example, exponentials, are inefficient. In such cases, it is usually easier to implement a look-up table with polynomial interpolation. Such a pipeline would already be sufficient for sum-and-threshold networks, which self-evidently are a (possibly non-polynomial) function upon a polynomial. It also adequately covers the right-hand side of differential equations: thus, to solve such equations, it remains to pass them into a solver. For very simple cases, it may be possible to solve them analytically, but for the general case, Euler-method evaluation using a programmable time step appears to be adequate.

The combination of design styles we used within a concurrent system of heterogeneous components is very similar to typical situations encountered in neural network modelling. This synergy drove our adoption of the hardware design flow as a model for neural application development. Notably, the environment we are implementing incorporates similar principles: off-the-shelf component reuse through standard neural library components; synthesis-directed network configuration using automated tools to generate the hardware mapping; and mix of design abstraction levels. The essential feature of hardware design systems: native support for concurrent description and simulation, is likewise essential in real-time neural network modelling, and we also note, *not* essential in ordinary software development. In particular, most software development does not incorporate an intrinsic notion of time: the algorithm deterministically sets the process flow which then proceeds as fast as the CPU will allow. Hardware design systems, by contrast, must of necessity include a notion of time, given that timing verification is one of the most important parts of the process. HDL-like systems also provide a logical evolutionary migration path from software simulation to hardware implementation, since the same model, with different library files, can be used to simulate at a high level, to develop hardware systems, or to instantiate a model onto a developed hardware platform.

Considerable work remains to be done both on SpiNNaker and generally in the area of neural development tools. SpiNNaker hardware is now available, making testing the models on the physical hardware an obvious priority. Very preliminary results successfully ran a simplified version of the "doughnut hunter" application, however, there is clearly a great deal more hardware testing to do. The increased speed will allow tests with larger and more complex models. We are currently working on implementing larger-scale, more biologically realistic models that simulate major subsystems of the brain and are scalable across a wide range of model sizes. For *very* large systems statistical description models as well as formal theories for neural network model design may be necessary. A second chip, containing more processors and larger routing flexibility is about to be sent for manufacture. There remains an open question of verification in a GALS system such as SpiNNaker: with nondeterministic timing behaviour, exact replication of the output from simulation is both impossible and irrelevant. It is possible to envision a "characterisation" stage where the developer compares SpiNNaker results against a variety of models with different parameters. SpiNNaker's ability to reconfigure dynamically on the fly becomes a significant advantage in this scenario.

With respect to the software model, developing a high-level mapping tool [15] to automate the "synthesis" process is a priority. We are also working on creating a user development environment based on the emerging PyNN neural description standard [29] that allows the modeller to implement the neural model with a high-level graphical or text description and use the automated generation flow to instantiate it upon SpiNNaker. Work is ongoing on extending the neural library with additional neural models, notably to extend the leaky-integrate-and-fire neuron with voltage-gated NMDA synapses. The NMDA synapse adds a second, "slow" time domain to the model in addition to the "fast" one of the neural dynamic and, like the "timeless" MLP model, extends the capabilities of the system towards a truly general-purpose neural modelling environment.

8.7 Conclusions

Design of an integrated hardware/software system like SpiNNaker provides a powerful model for neural network simulation: the hardware design flow of behavioural description, system synthesis, and concurrent simulation. With this work, we also emphasize one of the most important features of our hardware-design-flow methodology: the ability to leverage existing industry-standard tools and simulators so that it is unnecessary to develop a complete system from the ground up. The concept SpiNNaker embodies: that of a plastic hardware device incorporating dedicated neural components but neither hardwired to a specific model nor an entirely general-purpose reconfigurable device such as an FPGA, is the hardware equivalent of the software model, and is a new and perhaps more accessible neural hardware architecture than previous model-specific designs.

By making the hardware platform user-configurable rather than fixed-model, we introduce a new type of neural device whose architecture matches the requirements of large-scale experimental simulation, where the need to configure and test multiple and potentially heterogeneous neural network models within the same environment is critical. Such a model, we propose, is more suitable to neural network modelling than existing systems, especially when the model has asynchronous real-time dynamics. An asynchronous event-driven communications model makes it easier to design for arbitrary model timing and delays since it is not necessary to sample updates according to a global clock. This is both more biologically realistic and more representative of true parallel computing.

The pre-eminent feature of the software model, characteristic of native parallel computation, is *modularisation of dependencies*. This includes not only *data* dependencies (arguably, the usual interpretation of the term), but also temporal and abstractional ones. In other words, the model does not place restrictions on execution order between modules, or on functional support between different levels of software and hardware abstraction. Architecturally, the 3 levels of software abstraction distribute the design considerations between different classes of service and allow a service in one level to ignore the requirements of another, so that, for example, a Model level neuron can describe its behaviour without having to consider how or even if a System level service implements it. Structurally, it means that services operate independently and ignore what may be happening in other services, which from their point of view happen "in another universe" and only communicate via events "dropping from the sky", so to speak. Such a model accurately reflects the true nature of parallel computing and stands in contrast to conventional parallel systems that require coherence checking or coordination between processes.

Exploration of nondeterministic time effects also seems likely to occupy a growing interest within the neural research community, and devices such as SpiNNaker that have similar properties could reveal behaviours unobservable in conventional processors. In time it would also be ideal to move from a GALS system to a fully asynchronous system, ultimately, perhaps, to incorporate analogue neuromorphic components. Such a hybrid system would offer configurable processing with the speed and accuracy of analogue circuits where appropriate, instantiatable using a descendant of the software model we have developed. In that context, the function pipeline model we developed may be a useful abstraction for neural hardware, regardless of platform. To create the function pipeline, we attempted to decompose the general form of neurodynamic state equations into platform-neutral components that hardware can typically implement easily. Digital hardware can readily implement memories to form variable retrieval and LUT stages, and both analogue and digital hardware have effective blocks for polynomial evaluation and interpolation. Both DSPs and various analogue blocks offer efficient computation of differential equations. Thus, one could build a neural system in building-block fashion, by chaining together various components using AER signalling, allowing for the construction of hybrid systems in addition to integrated approaches like SpiNNaker. This system is the future version, as much as SpiNNaker is the present version, of a neural network matching the development model and environment to the computational model.

Acknowledgements The Spinnaker project is supported by the Engineering and Physical Sciences Research Council, partly through the Advanced Processor Technologies Platform Partnership at the University of Manchester, and also by ARM and Silistix. Steve Furber holds a Royal Society-Wolfson Research Merit Award.

References

1. Jahnke A, Roth U, Schönauer T (1999) Digital simulation of spiking neural networks. In: Pulsed neural networks. MIT Press, Cambridge, pp 237–257
2. Izhikevich E, Edelman GM (2008) Large-scale model of mammalian thalamocortical systems. Proc Natl Acad Sci USA 105(9):3593–3598
3. Westerman WC, Northmore DPM, Elias JG (1999) Antidromic spikes drive hebbian learning in an artificial dendritic tree. Analog Circuits Signal Process 19(2–3):141–152
4. Mehrtash N, Jung D, Hellmich HH, Schönauer T, Lu VT, Klar H (2003) Synaptic plasticity in spiking neural networks (SP^2INN) a system approach. IEEE Trans Neural Netw 14(5):980–992
5. Indiveri G, Chicca E, Douglas R (2006) A VLSI array of low-power spiking neurons and bistable synapses with spike-timing dependent plasticity. IEEE Trans Neural Netw 17(1):211–221
6. Furber SB, Temple S, Brown A (2006) On-chip and inter-chip networks for modelling large-scale neural systems. In: Proc international symposium on circuits and systems, ISCAS-2006, May

7. Oster M, Whatley AM, Liu SC, Douglas RJ (2005) A hardware/software framework for real-time spiking systems. In: Proc 15th int'l conf artificial neural networks (ICANN2005). Springer, Berlin, pp 161–166
8. Johansson C, Lansner A (2007) Towards cortex sized artificial neural systems. Neural Netw 20(1):48–61
9. Khan MM, Lester D, Plana L, Rast A, Jin X, Painkras E, Furber S (2008) SpiNNaker: Mapping neural networks onto a massively-parallel chip multiprocessor. In: Proc 2008 int'l joint conf on neural networks (IJCNN2008)
10. Plana LA, Furber SB, Temple S, Khan MM, Shi Y, Wu J, Yang S (2007) A GALS infrastructure for a massively parallel multiprocessor. IEEE Des Test Comput 24(5):454–463
11. Goldberg D, Cauwenberghs G, Andreou A (2001) Analog VLSI spiking neural network with address domain probabilistic synapses. In: Proc 2001 IEEE int'l symp circuits and systems (ISCAS2001). IEEE Press, New York, pp 241–244
12. Izhikevich E (2003) Simple model of spiking neurons. IEEE Trans Neural Netw 14:1569–1572
13. Rast A, Yang S, Khan MM, Furber S (2008) Virtual synaptic interconnect using an asynchronous network-on-chip. In: Proc 2008 int'l joint conf on neural networks (IJCNN2008)
14. ARM Limited (2002) ARM9E-S technical reference manual
15. Brown A, Lester D, Plana L, Furber S, Wilson P (2009) SpiNNaker: the design automation problem. In: Proc 2008 int'l conf neural information processing (ICONIP 2008). Springer, Berlin
16. Lettnin D, Braun A, Bodgan M, Gerlach J, Rosenstiel W (2004) Synthesis of embedded SystemC design: a case study of digital neural networks. In: Proc design, automation & test in Europe conf & exhibition (DATE'04), vol 3, pp 248–253
17. Modi S, Wilson P, Brown A, Chad J (2004) Behavioral simulation of biological neuron systems in SystemC. In: Proc 2004 IEEE int'l behavioral modeling and simulation conf, pp 31–36
18. Panda P (2001) SystemC—a modeling platform supporting multiple design abstractions. In: Proc int'l symp on system synthesis
19. Rast A, Jin X, Khan M, Furber S (2009) The deferred-event model for hardware-oriented spiking neural networks. In: Proc 2008 int'l conf neural information processing (ICONIP 2008). Springer, Berlin
20. Sloss AN, Symes D, Wright C (2004) ARM system developer's guide—designing and optimizing system software. Morgan Kaufmann, San Francisco
21. Khan MM, Jin X, Furber S, Plana L (2007) System-level model for a GALS massively parallel multiprocessor. In: Proc 19th UK asynchronous forum, pp 9–12
22. Wang HP, Chicca E, Indiveri G, Sejnowski TJ (2008) Reliable computation in noisy backgrounds using real-time neuromorphic hardware. In: Proc 2007 IEEE biomedical circuits and systems conf (BIOCAS2007), pp 71–74
23. Daud T, Duong T, Tran M, Langenbacher H, Thakoor A (1995) High resolution synaptic weights and hardware-in-the-loop learning. Proc SPIE 2424:489–500
24. Jin X, Furber S, Woods J (2008) Efficient modelling of spiking neural networks on a scalable chip multiprocessor. In: Proc 2008 int'l joint conf on neural networks (IJCNN2008)
25. Izhikevich E (2006) Polychronization: computation with spikes. Neural Comput 18(2)
26. Jin X, Rast A, Galluppi F, Khan M, Furber S (2009) Implementing learning on the spinnaker universal chip multiprocessor. In: Proc 2009 int'l conf neural information processing (ICONIP 2009). Springer, Berlin
27. Bi G, Poo M (1998) Synaptic modifications in cultured hippocampal neurons: dependence on spike timing, synaptic strength, and postsynaptic cell type. J Neurosci 18(24):10464–10472
28. Tiesenga PHE, Sejnowski TJ (2001) Precision of pulse-coupled networks of integrate-and-fire neurons. Network 12(2):215–233
29. Davidson AP, Brüderle D, Eppler J, Kremkow J, Muller E, Pecevski D, Perrinet L, Yger P (2009) PyNN: a common interface for neuronal network simulators. Front Neuroinf 2(11)

Chapter 9
Solving Complex Control Tasks via Simple Rule(s): Using Chaotic Dynamics in a Recurrent Neural Network Model

Yongtao Li and Shigetoshi Nara

Abstract The discovery of chaos in the brain has suggested that chaos could be essential to brain functioning. However, a challenging question is what is the role of chaos in brain functioning. As for our endeavor on this question, the investigation on the functional aspects of chaotic dynamics from applicable perspective is emphasized. In this chapter, chaotic dynamics are introduced in an quasi-layered recurrent neural network model (QLRNNM), which incorporates sensory neurons for sensory information processing and motor neurons for complex motion generation. In QLRNNM, two typical properties of chaos are utilized. One is the sensitive response to external signals. The other is the complex dynamics of many but finite degrees of freedom in a high dimensional state space, which can be utilized to generate low dimensional complex motions by a simple coding. Moreover, inspired by neurobiology, presynaptic inhibition is introduced to produce adaptive behaviors. By virtue of these properties, a simple control algorithm is proposed to solve two-dimensional mazes, which is set as an ill-posed problem. The results of computer experiments and actual hardware implementation show that chaotic dynamics emerging in sensory neurons and motor neurons enable the robot to find complex detours to avoid obstacles via simple control rules. Therefore, it can be concluded that the use of chaotic dynamics has novel potential for complex control via simple control rules.

9.1 Introduction

With the rapid progress of modern science and technology, knowledge of the structure and functionality of biological systems including the brain has developed considerably in widespread areas of research. However, the present situation is still filled with complexity. Due to the giant variety of interactions among a great number of elements existing in systems, enormously complex nonlinear dynamics emerging

Y. Li (✉) · S. Nara
Department of Electrical & Electronic Engineering, Graduate School of Natural Science and Technology, Okayama University, Okayama, Japan
e-mail: yongtaoli@es.hokudai.ac.jp

S. Nara
e-mail: nara@chaos.elec.okayama-u.ac.jp

in them make their mechanisms difficult to understand. The conventional methodology based on reductionism prefers to decompose a complex system into more and more individual parts until every part is understood, so it may more or less fall into the following two difficulties: *combinatorial explosion* and *divergence of algorithmic complexity*. On the other hand, recent innovative research in biological information and control processing, particularly brain functions, has suggested that they could work with novel *dynamical mechanisms* that realize complex information processing and/or adaptive controlling [1, 3–6, 9, 14, 21, 22]. Our core idea is to harness complex nonlinear dynamics in information processing and control. If complex nonlinear dynamics could be applied to realizing complex information processing and control, it could be useful for the understanding of mechanisms in biological systems in general, and brains in particular. This is our primary motivation for studying chaotic dynamics in neural networks from heuristic functional viewpoints.

We think that harnessing the onset of chaotic dynamics in neural networks could be applied to realizing complex control or complex information processing with simple control rules. Along this idea, Nara and Davis introduced chaotic dynamics in a recurrent neural network model (RNNM) consisting of binary neurons, in which the onset of chaotic dynamics could be controlled by means of changing a system parameter—*connectivity*. Using a simple rule based on adaptive switching of connectivity, chaotic dynamics was applied to solving complex problems, such as solving a memory search task which is set in an ill-posed context [14]. In these computer experiments, it was shown that chaotic dynamics observed in these systems could have important potential for complex information processing and controlling in a particular context, where the environment gives the systems simple responses. "Non-specific" responses result in useful complex dynamics that were called "constrained chaos" [13]. The different aspect of these dynamics was also discussed as concerning "chaotic itinerancy" [8, 22]. Later, chaotic dynamics in RNNM was applied to adaptive control tasks with ill-posed properties. Computer experiments showed that a roving robot could successfully solve two-dimensional mazes [20], or capture a moving target [11] using chaotic dynamics in RNNM using simple control rules. These studies not only indicated that chaotic dynamics could play very important roles in complex control via simple control rules, but also showed how to use the chaotic dynamics in a general way. However, comparing with biological neural systems, it can be said that the previous works could be improved in the following two points. First, sensory neuron systems were not considered. Second, the adaptive switching of connectivity was an unnatural control method which is not observed in real biological systems.

Motivated by these points, we have proposed a QLRNNM consisting of sensory neurons and motor neurons. In previous works, Mikami and Nara found that chaos has a property of sensitive response to external input [12]. In this chapter, chaotic dynamics were introduced into both sensory neurons and motor neurons. Two typical properties of chaos are utilized. One is sensitive response to external signals in sensory neurons and that of motor neurons to input signals from sensory neurons. The other is complex dynamics of many but finite degrees of freedom in high dimensional state space, which can be utilized to generate low dimensional complex motions by a simple coding. So this model is robust and has large redundancy.

9 Solving Complex Control Tasks via Simple Rule(s) 161

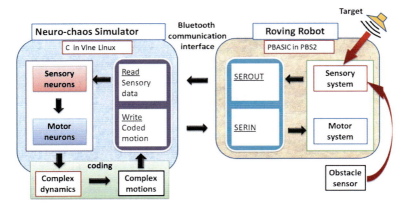

Fig. 9.1 Block diagram of control system

Moreover, a neurobiology-inspired method, presynaptic inhibition, is introduced to produce adaptive behavior for solving complex problems. Using these properties, chaotic dynamics in QLRNNM is applied to solving ill-posed problems, in particular, auditory behavior of animals. More specifically, a female cricket can track the direction of the position of a male by detecting the calling song of the male [7]. In dark fields with a large number of obstacles, this auditory behavior of crickets includes two ill-posed properties. One is that darkness and noise in the environment prevent a female from accurately deciding the directions of a male, and the other is that a large number of big obstacles in fields force females to solve two-dimensional mazes. Therefore, using chaotic dynamics in QLRNNM, a simple control algorithm is proposed for solving a two-dimensional maze, which is set as an ill-posed problem. The results of computer experiments show that chaotic dynamics in QLRNNM have potential to solve this typical ill-posed problem via simple control rules.

Furthermore, we have constructed a roving robot, regarded as a metaphor model of a female cricket, to solve mazes in which there is a target emitting a specified sound like a male cricket. In the case of only motor neurons, a preliminary experiment with the hardware implementation shows that the roving robot could successfully find an appropriate detour to avoid obstacles and reach the target using chaotic dynamics with a simple control rule, as has been reported in [10]. In the present stage, sensing neurons have been considered and implemented into the robot. Experiments with the hardware implementation indicates that the robot can solve two-dimensional mazes using chaotic dynamics which emerge in both sensory and motor neurons.

9.2 Construction of Control System

Our control system consists of four main components (target, robot, communication interface and neuro-chaos simulator), which are shown in Fig. 9.1.

Fig. 9.2 The roving robot with sensors including two ultrasonic sensors and four microphones: photograph (*left*) and schematic diagram (*right*)

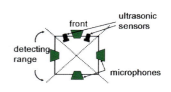

In our study, a roving robot has been developed from the robot proposed in [10]. It has been constructed using a Boe-Bot robot with two driving wheels and one castor (2DW1C) from Parallax company, as shown in Fig. 9.2. The MPU of the robot is a BASIC Stamp 2 programmable microcontroller (PBS2) whose code is written in the language PBASIC. A system of six sensors which can be divided into two sub-systems has been designed and installed on the robot. One sub-system is the sensing system for detecting obstacles that consists of a pair of ultrasonic sensors, which can provide real-time signals about the existence of obstacles in front of the robot. The other is the sensing system for detecting sound signals from the target, which consists of four sets of directional microphone circuits which function as the *ears* of the robot. Four directional microphones are set in directions to the front, the back, the left and the right of the robot. Each directional microphone consists of a condenser microphone and a parabolic reflector made of a parabola-shaped plastic dish. The measure of its directivity (i.e., dependence of received signal strength on direction) gives a normal distribution.

In our study, a loudspeaker is employed as the target, emitting a 3.6 KHz sound signal like a singing cricket. The sound signal from the target is picked up by those four directional microphones, and forms four channels of sound signal with different intensity. And then, these four separate sound signals are respectively amplified, rectified by the internally equipped electronic circuits, digitalized by a A/D converter (ADC0809), and transferred to PBS2. The data from the sensors is sent from PBS2 to the neuro-chaos simulator via a Bluetooth communication interface and input into the corresponding subset of sensory neurons. Responding sensitively to external input, chaotic dynamics emerging in sensory neurons causes adaptive dynamics in motor neurons, which are transformed into adaptive motion by a particular coding method. Finally, motion signals sent to PBS2 make the robot move adaptively. Due to poor performance of the robot MPU, a computer (Pentium III CPU 1 GHz and 1 GB memory) is employed as the neuro-chaos simulator, which is programmed in C language and works in Vine Linux. So, a Bluetooth communication interface is designed to provide wireless communication between the roving robot and the neuro-chaos simulator. Once the hardware system is ready, the most important issue is how the neuro-simulator works, which will be mentioned later in detail. The setting for the task of solving mazes in this study is as follows.

(1) A two-dimensional maze is setup
(2) The robot has no pre-knowledge about obstacles
(3) The target emits a sound signal

(4) The robot acquires sensory information
- It detects whether or not there are obstacles preventing it from moving forward, using the ultrasonic sensors
- It detects the rough direction of the target using the four microphones

(5) The robot calculates movement increments corresponding to every time step of neural network activity

We comment that in the above setting, there are two ill-posed problems. First, the robot has no pre-knowledge about obstacles. This means that the robot has to work autonomously without need of human intervention, so the existence of solution is not guaranteed. Second, the robot only obtains rough target information instead of accurate orientation, so the uniqueness of solution is not guaranteed even if the robot can reach target, as means the possibility of multiple detours. This setting is similar to the context for biological behavior.

9.3 From RNNM to QLRNNM

The neuro-chaos simulator, a "brain-morphic device" on the robot, is utilized to implement dynamical activities of a neural network. Here, we give a brief description of the chaotic dynamics in the neural network.

9.3.1 Recurrent Neural Network Model (RNNM)

We work on a recurrent neural network model, the updating rule of which is defined by

$$S_i(t+1) = \text{sgn}\left(\sum_{j \in G_i(r)} W_{ij} S_j(t)\right),$$

$$\text{sgn}(u) = \begin{cases} +1, & u \geq 0; \\ -1, & u < 0, \end{cases}$$

(9.1)

where $S_i(t) = \pm 1$ ($i = 1 \sim N$) represents the firing state of the i-th neuron at time t. W_{ij} is an asymmetrical synaptic weight from the neuron S_j to the neuron S_i, where W_{ii} is taken to be 0. $G_i(r)$ means a spatial configuration set of connectivity r ($0 < r < N$), that is fan-in number, for the neuron S_i. At a certain time t, the firing state of neurons in the network can be represented as a N-dimensional state vector $\mathbf{S}(t)$, called a state pattern. The updating rule shows that time development of state pattern $\mathbf{S}(t)$ is determined by two factors—the synaptic weight matrix $\{W_{ij}\}$ and connectivity r. Therefore, when full connectivity $r = N - 1$ is employed, W_{ij} can be determined so as to embed an associative memory into the network. In other words, once W_{ij} is determined appropriately, an arbitrarily chosen state pattern $\mathbf{S}(t)$

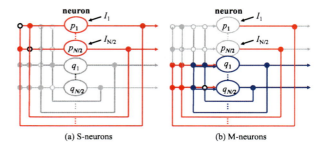

Fig. 9.3 Quasi-layered RNNM

could evolve into one of multiple stationary states as time evolves. In our study, a kind of orthogonalized learning method [2] is utilized to determine W_{ij}, which is defined by

$$W_{ij} = \sum_{\mu=1}^{L} \sum_{\lambda=1}^{K} (\xi_\mu^{\lambda+1})_i \cdot (\xi_\mu^\lambda)_j^\dagger, \tag{9.2}$$

where $\{\xi_\mu^\lambda \mid \lambda = 1, \ldots, K, \mu = 1, \ldots, L\}$ is an attractor pattern set, K is the number of memory patterns included in a cycle and L is the number of memory cycles. $\xi_\mu^{\lambda\dagger}$ is the conjugate vector of ξ_μ^λ which satisfies $\xi_\mu^{\lambda\dagger} \cdot \xi_{\mu'}^{\lambda'} = \delta_{\mu\mu'} \cdot \delta_{\lambda\lambda'}$, where δ is Kronecker's delta. This method was confirmed to be effective to avoid spurious attractors that affect L attractors with K-step maps embedded in the network when connectivity $r = N$ [13–17, 19].

9.3.2 QLRNNM

A QLRNNM is an asymmetrical RNNM with quasi-layered structure which has N binary neurons consisting of an upper layer with $N/2$ sensory neurons (S-neurons) and a lower layer with $N/2$ motor neurons (M-neurons). An S-neuron has only self-recurrence and receive external input signals, as shown in Fig. 9.3(a). On the other hand, an M-neuron has both self-recurrence and recurrent outputs from S-neurons, as shown in Fig. 9.3(b).

At time t, the firing state of neurons in QLRNNM can be represented by a N dimensional state vector $\mathbf{S}(t)$, called a state pattern. $\mathbf{S}(t) = [\mathbf{p}(t), \mathbf{q}(t)]$, where $\mathbf{p}(t) = \{p_i(t) = \pm 1 \mid i = 1, 2, \ldots, N/2\}$ and $\mathbf{q}(t) = \{q_i(t) = \pm 1 \mid i = 1, 2, \ldots, N/2\}$. The updating rules of S-neurons and M-neurons are defined by

S-neurons:

$$\begin{cases} p_i(t+1) = \text{sgn}(\sum_{j \in G(r_{u,u})} W_{ij}^{u,u} p_j(t) + \varphi_i(t)), \\ \varphi_i(t) = \alpha_i(t) I_i(t) \quad (i \in F_k(l)). \end{cases} \tag{9.3}$$

9 Solving Complex Control Tasks via Simple Rule(s)

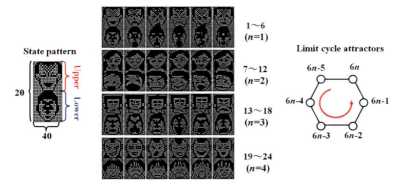

Fig. 9.4 One example of embedded attractor patterns

M-neurons:

$$\begin{cases} q_i(t+1) = \text{sgn}(\sum_{j \in G(r_{l,l})} W_{ij}^{l,l} q_j(t) + \psi_i(t)), \\ \psi_i(t) = \sum_{j \in G(r_{u,l})} W_{ij}^{u,l} p_j(t), \end{cases} \quad (9.4)$$

where $W_{ij}^{u,l}$ is a connection weight from neuron p_j of S-neurons to neuron q_i of M-neurons, and $W_{ij}^{u,u}$ and $W_{ij}^{l,l}$ are defined similarly. In the case of multisensory input (for example, up to P sensors), $F_k(l)$ is a subset with l members of S-neurons ($I_i(t) = \pm 1$) corresponding to one sensor k ($1 < k < P$). $\alpha_i(t)$ is the strength of external input depending on the subset $F_k(l)$. If we do not consider external input signals, that is, $\alpha_i(t) = 0$, the updating rules show that time development of state pattern $\mathbf{S}(t)$ depends on two inherent system parameters—connection weight matrix W_{ij} and connectivity r. Therefore, the orthogonalized learning method in RNNM can also be used to determine connection weight matrix $\mathbf{W} = \{W_{ij}\}$ so as to embed cyclic memory attractors in N dimensional state space. For the quasi-layered structure of QLRNNM, matrix $\mathbf{W}$ can be divided into four ($N/2 \times N/2$) dimensional sub-matrices, represented by

$$\mathbf{W} = \begin{bmatrix} \mathbf{W}^{u,u} & \mathbf{W}^{l,u} \\ \mathbf{W}^{u,l} & \mathbf{W}^{l,l} \end{bmatrix} = \begin{bmatrix} \mathbf{W}^{u,u} & \mathbf{0} \\ \mathbf{W}^{u,l} & \mathbf{W}^{l,l} \end{bmatrix}, \quad (9.5)$$

here $\mathbf{W}^{u,u} = \{W_{ij}^{u,u}\}$, $\mathbf{W}^{u,l} = \{W_{ij}^{u,l}\}$ and $\mathbf{W}^{l,l} = \{W_{ij}^{l,l}\}$. Since S-neurons have no feedback from M-neurons, $\mathbf{W}^{l,u}$ is a ($N/2 \times N/2$) dimensional zero matrix. Using the orthogonalized learning method in (9.2) to determine W_{ij}, ($L \times K$) state patterns can be embedded as L cyclic memory attractors with period K. For example, in Fig. 9.4, 24 ($K = 6$ and $L = 4$) state patterns consisting of $N = 800$ neurons are embedded into QLRNNM.

In the case of sufficiently large connectivity r ($r_{u,u} \cong N/2$, $r_{u,l} \cong N/2$, $r_{l,l} \cong N/2$), as the network evolves with the updating rules shown in (9.3) and (9.4), after enough time steps, any randomly initial state pattern will converge into one of the embedded cyclic attractors, that is, pattern periodic behavior with period of K

steps emerges in the output sequence of state. In this sense, when connection weight matrix $\{W_{ij}\}$ is appropriately determined by the orthogonalized learning method, in the case of large enough connectivity r ($r_{u,u}$, $r_{u,l}$, $r_{l,l}$), the network functions as a conventional associative memory.

9.3.3 Introducing Chaotic Dynamics in QLRNNM

As described in previous works, connectivity r is the key system parameter for introducing chaotic dynamics in RNNM. If connectivity r is reduced by blocking signal propagation among neurons, so connectivity r becomes smaller and smaller, memory attractors gradually become deformed. Finally, attractors become unstable in N dimensional state space and chaotic wandering occurs in high dimensional state space [13]. In order to investigate the inherent properties of QLRNNM, we could omit the external input signals, that is, $\alpha_i(t) = 0$. Then, after weight matrix $\mathbf{W}$ has been determined, connectivity r is also the key system parameter for introducing chaotic dynamics in QLRNNM. Due to the special structure of QLRNNM, three kinds of connectivity ($r_{u,u}$, $r_{u,l}$, $r_{l,l}$) have direct effect on time development of state patterns. So chaotic dynamics in QLRNNM have various characteristic properties depending on choices of these parameter values.

Generally speaking, if chaotic dynamics could be observed in a system, it usually shows obvious bifurcation depending on some system parameters. Therefore, a method to analyze the destabilizing process is employed to illustrate the bifurcation of QLRNNM. One-dimensional projection of N dimensional state pattern $\mathbf{S}(t)$ to a certain N dimensional reference pattern $\mathbf{S}(0)$ is used to define "bifurcation overlap" $m(t)$, which is defined by

$$m(t) = \frac{1}{N}\mathbf{S}(0) \cdot \mathbf{S}(t), \qquad (9.6)$$

$$t = Kl + t_0 \quad (l = 1, 2, \ldots), \qquad (9.7)$$

where $\mathbf{S}(0)$ is an initial pattern(reference pattern) and $\mathbf{S}(t)$ is the state pattern at time step t. The overlap $m(t)$ is a normalized inner product, so $-1 \le m(t) \le 1$. If $m(t) = 1$, the state pattern $\mathbf{S}(t)$ and the reference pattern $\mathbf{S}(0)$ are the same pattern. Therefore, $m(t) \equiv 1$ means that the reference pattern $\mathbf{S}(0)$ periodically appears every K steps in the evolving sequence of state pattern $\mathbf{S}(t)$. Otherwise, $m(t)$ is not constant and is distributed near 0, as means that state pattern $\mathbf{S}(t)$ evolves non-periodically.

Now let us describe chaotic dynamics in QLRNNM using a bifurcation diagram. First, $N/2$ S-neurons has only self-recurrence, so reducing connectivity $r_{u,u}$ could introduce chaotic dynamics into S-neurons. Referring to (9.6), the bifurcation overlap of S-neurons can be defined by

$$m_u(t) = \frac{2}{N}\mathbf{p}(0) \cdot \mathbf{p}(t). \qquad (9.8)$$

9 Solving Complex Control Tasks via Simple Rule(s)

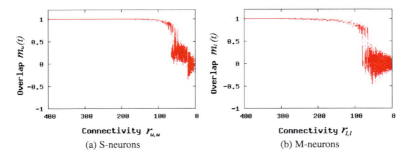

Fig. 9.5 The long-time behaviors of overlap $m_u(t)$ and $m_l(t)$ at K-step mappings: The *horizontal axis* (**a**) is $r_{u,u}$ (0–399); (**b**) is $r_{l,l}$ (0–399) in the case of $r_{u,l} = 0$

The overlap $m_u(t)$ with respect to $r_{u,u}$ is shown in Fig. 9.5(a). When $r_{u,u}$ is sufficiently large, embedded cyclic attractors in S-neurons are stable in $N/2$ dimensional state space, which results in $m_u(t) \equiv 1$. With the decrease of $r_{u,u}$, attractors gradually become deformed, so $m_u(t) \leq 1$. Finally, when $r_{u,u}$ becomes quite small, attractors become unstable and chaotic wandering emerges in S-neurons, so $m_u(t)$ shows a strong overlap.

Next, let us introduce the overlap $m_l(t)$ of M-neurons, which is defined by

$$m_l(t) = \frac{2}{N}\mathbf{q}(0) \cdot \mathbf{q}(t). \qquad (9.9)$$

In QLRNNM, M-neurons are affected not only by self-recurrence but also by recurrent output of S-neurons, so three kinds of connectivity ($r_{u,u}$, $r_{u,l}$ and $r_{l,l}$) act on M-neurons together. In order to show the characteristic properties of chaotic dynamics in QLRNNM, their effects on dynamical behavior emerging in M-neurons are considered respectively. First, we take $r_{u,l} = 0$, that is, S-neurons can not affect M-neurons. Under this condition, the overlap $m_l(t)$ of M-neurons with respect to $r_{l,l}$ is shown in Fig. 9.5(b). It indicates that, without the effect of S-neurons, reducing $r_{l,l}$ also introduces chaotic dynamics in M-neurons, as in S-neurons. Second, if we take $r_{u,l} \neq 0$, M-neurons are affected by both S-neurons and M-neurons. In the case of $r_{u,l} = 400$, when $r_{u,u}$ is a larger connectivity, quite small $r_{l,l}$ can not introduce chaotic dynamics into M-neurons, as is shown in Fig. 9.6(a). On the other hand, once $r_{u,u}$ is a sufficiently small connectivity capable of introducing chaotic dynamics in S-neurons, even if $r_{l,l}$ is still large enough, chaotic dynamics also emerges in M-neurons, as is shown in Fig. 9.6(b). These results indicate that M-neurons could sensitively (strongly) respond to S-neurons.

In order to investigate the extent of sensitivity of response, we have investigated the overlap $m_u(t)$ of S-neurons and $m_l(t)$ of M-neurons along time axis in the case of only changing $r_{u,l}$. In Fig. 9.7(a), chaotic dynamics emerging in S-neurons can not affect M-neurons due to $r_{u,l} = 0$. However, Fig. 9.7(c) and Fig. 9.7(d) show that, chaotic dynamics in M-neurons is stronger than that in S-neurons. Therefore, these figures also indicate that, M-neurons respond more sensitively to S-neurons with the increase of $r_{u,l}$. Particularly, in Fig. 9.7(d), when the overlap $m_u(t)$ of S-neurons approaches 1 a little, M-neurons show quick response and the overlap $m_l(t)$

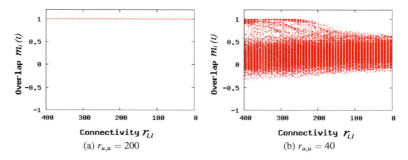

Fig. 9.6 The long-time behaviors of overlap $m_l(t)$ at K-step mappings in the case of $r_{u,l} = 400$: The *horizontal axis* is $r_{l,l}$ (0–399)

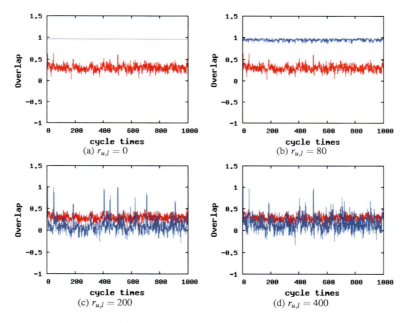

Fig. 9.7 The overlaps $m_u(t)$ (*red line*) and $m_l(t)$ (*blue line*) along the time axis. $r_{u,u} = 40$ and $r_{l,l} = 200$. The *horizontal axis* represents the p-th K-steps, where $t = Kp + t_0$ ($K = 6$ and $t_0 = 1200$)

of M-neurons approaches 1 immediately. When chaotic dynamics were introduced in S-neurons, stronger chaotic dynamics emerge in M-neurons. Concerning the sensitive response property of chaos to external input [12], a novel idea arises. It may be possible for M-neurons to produce adaptive dynamics to control the robot due to their sensitive response to S-neurons which respond to external input. This idea will be used in the control algorithm described later.

9 Solving Complex Control Tasks via Simple Rule(s)   169

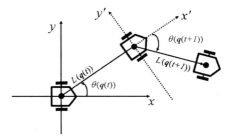

Fig. 9.8 The motion of the robot: the robot has local coordinates in which x axis always corresponds to the front of the robot

9.4 A Simple Coding Method for Control

9.4.1 Motion Functions

As has been described in the above section, depending on the three kinds of connectivity ($r_{u,u}$, $r_{u,l}$ and $r_{l,l}$) in QLRNNM complex dynamics can emerge in S-neurons and M-neurons. How could the complex dynamical behavior be utilized to realize motion control? Referring to biological data that several tens of muscles to maintain usual motions are controlled by the activities of about 10 billion ($\sim 10^{10}$) parietal-lobe neurons in mammal brains, it is considerably reasonable that, by virtue of a certain appropriate coding, complex dynamics generated by large numbers of neurons are transformed into various motions with a smaller numbers of degrees of freedom. In this study, since the robot has two driving wheels and one castor (2DW1C), its motion includes two steps as shown in Fig. 9.8. First, it rotates an angle $\theta(t)$ and then moves forward a distance $L(t)$. Therefore, in order to realize 2-dimensional motion of the robot using complex dynamics emerging in M-neurons in QLRNNM, the $N/2$ dimensional state pattern $\mathbf{q}(t)$ is transformed into the rotation angle $\theta(\mathbf{q}(t))$ and the movement distance $L(\mathbf{q}(t))$ by simple coding functions, called *motion functions* and defined by

$$\theta(\mathbf{q}(t)) = \frac{\pi}{2}\left(\frac{\mathbf{A}\cdot\mathbf{B}}{N/4} - \frac{\mathbf{C}\cdot\mathbf{D}}{N/4}\right), \qquad (9.10)$$

$$L(\mathbf{q}(t)) = \frac{\pi d}{\sqrt{2}}\sqrt{\left(\frac{\mathbf{A}\cdot\mathbf{B}}{N/4}\right)^2 + \left(\frac{\mathbf{C}\cdot\mathbf{D}}{N/4}\right)^2}, \qquad (9.11)$$

where $\mathbf{A}$, $\mathbf{B}$, $\mathbf{C}$, $\mathbf{D}$ are four $N/4$ dimensional sub-space vectors of state pattern $\mathbf{q}(t)$, which is shown in Fig. 9.10. Normalizing the inner products of $\mathbf{A}\cdot\mathbf{B}$ and $\mathbf{C}\cdot\mathbf{D}$ gives a value form -1 to 1. d is the diameter of the wheel. Therefore, the rotation angle $\theta(\mathbf{q}(t))$ takes value from $-\pi$ to π, and the movement distance $L(\mathbf{q}(t))$, from 0 to πd.

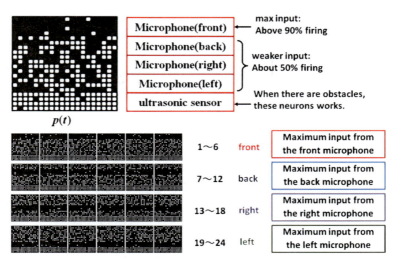

Fig. 9.9 Four attractors embedded in S-neurons

9.4.2 Design of Attractors for Controlling

9.4.2.1 Sensory Neurons p(t)

Sensory neurons **p**(*t*) consists of 5 independent sub-space vectors that correspond to 5 sensors—four microphones for detecting target direction and a pair of ultrasonic sensors regarded as one merged sensor for detecting obstacles, shown in Fig. 9.9. Four attractors are embedded which correspond to the cases in which there are no obstacles and the direction of the maximum signal intensity received by microphones is front, back, right, or left, respectively. S-neurons sensitively respond to external signal input, producing dynamics which affect M-neurons, which then produce dynamics which are transformed into motion. For example, in the case of no obstacle, if the maximum signal intensity is received by the front microphone, S-neurons sensitively respond to it and enable M-neurons to produce dynamics which can be transformed into forward motion by motion functions. On the other hand, once ultrasonic sensors find obstacles to prevent the robot from moving forward at any time, S-neurons corresponding to ultrasonic sensors become excited. The effect of the S-neurons on the M-neurons, which we think of as "presynaptic inhibition" as explained later, causes complex dynamics in M-neurons and enables the robot to find an appropriate detour to avoid the obstacle.

9.4.2.2 Motor Neurons q(t)

Four attractors which correspond to four simple prototype motions (forward, backward, leftward, rightward) are embedded in motor neurons in the lower layer, shown

9   Solving Complex Control Tasks via Simple Rule(s)   171

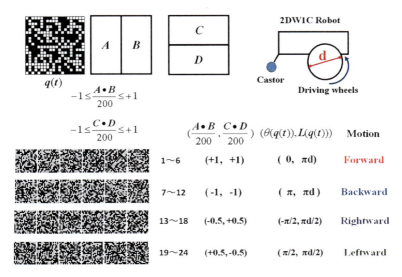

Fig. 9.10 Four attractors embedded in M-neurons: Each attractor corresponds to a prototypical simple motion

Fig. 9.11 Examples of motion control ($r_{u,l} = 400$): (*left*) $r_{u,u} = 200$ (*right*) $r_{u,u} = 30$

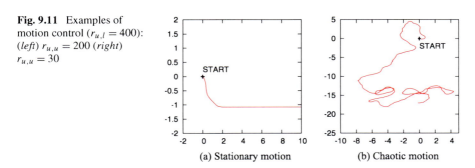

in Fig. 9.10. In this study, **A**, **B**, **C** and **D** are four $N/4$ dimensional sub-space vectors of state pattern **q**(t). Coding functions known as motion functions make the correspondence between attractors and four simple prototype motions in two-dimensional space.

9.4.3 Complex Motions

After attractors for controlling have been embedded into QLRNNM, if $r_{u,l}$ is sufficiently large so that M-neurons could sensitively respond to S-neurons, different $r_{u,u}$ could introduce various dynamics into M-neurons. Through the coding of motion functions, the robot shows complex motions. When $r_{u,u}$ is large enough, any initial state pattern converges into one of the embedded attractors as time evolves.

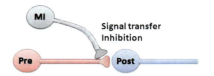

Fig. 9.12 Presynaptic inhibition: presynaptic neuron (*red*), postsynaptic neuron (*blue*) and modulatory interneuron (*gray*)

Correspondingly, the robot shows stationary motion after some steps. An example is shown in Fig. 9.11(a). On the other hand, if $r_{u,u}$ is quite small, chaotic dynamics emerge in S-neurons. This also causes chaotic dynamics in M-neurons. So the robot moves chaotically in two-dimensional space. An example is shown in Fig. 9.11(b).

9.5 A Simple Control Algorithm

Now it is time to discuss the control algorithm for solving two-dimensional mazes. Before we begin to talk about it, let us consider the auditory behavior of crickets, which is an ill-posed problem observed in a biological system. A female cricket has to avoid obstacles and find a way to reach a male cricket depending only on hearing the calling song of the male cricket. In the case of no obstacles, it responds directly to the calling song of the male cricket. However, when there are obstacles to prevent her from moving forward, it finds an appropriate detour to avoid them. We suppose that inhibition behaviors may weaken the direct response and produce adaptive motions that are useful for finding an appropriate detour to avoid obstacles. Therefore, an inhibition behavior similar to the *presynaptic inhibition* which is prevalent among biological neurons is introduced in our control algorithm for solving two-dimensional mazes.

As shown in Fig. 9.12, presynaptic inhibition means that, if modulatory interneurons fire, signal transfer from a presynaptic neuron to a postsynaptic neuron is blocked. In our algorithm, when the robot detects an obstacle, presynaptic inhibition is introduced to weaken the response of M-neurons to S-neurons. In our system, S-neurons for detecting obstacles are taken as n modulatory interneurons ($p_1^M \sim p_n^M$), S-neurons for receiving sound signals as m presynaptic neurons ($p_1^P \sim p_m^P$) and M-neurons as $N/2$ postsynaptic neurons ($q_1 \sim q_{N/2}$), as is shown in Fig. 9.13.

After presynaptic inhibition is introduced, the updating rule for M-neurons shown in (9.4) should be modified and redefined as

$$\begin{cases} q_i(t+1) = \mathrm{sgn}(\sum_{j \in G(r_{l,l})} W_{ij}^{l,l} q_j(t) + \psi_i(t)), \\ \psi_i(t) = \sum_{j, \beta \in G(r_{u,l})} W_{i\beta}^{u,l} \delta_{\beta j}^M p_j(t), \\ \delta_{\beta j}^M = (1 - p_k^M)/2, \end{cases} \quad (9.12)$$

where p_k^M represents the firing state of a modulatory interneuron with index k ($1 \leq k \leq n$) and $\{\delta_{\beta j}^M\}$ is a ($N/2 \times N/2$) diagonal matrix of presynaptic inhibition. If a modulatory interneuron becomes excited, presynaptic neurons which it acts on will experience presynaptic inhibition.

9 Solving Complex Control Tasks via Simple Rule(s)

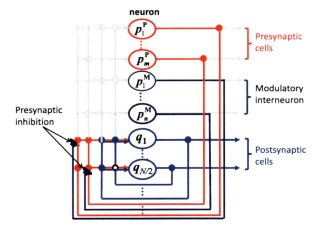

Fig. 9.13 Introducing presynaptic inhibition into QLRNNM

Exploiting the sensitivity of chaos in QLRNNM and presynaptic inhibition, a simple control algorithm is proposed for solving two-dimensional mazes, shown in Fig. 9.14. Where connectivity $r_{u,u}$ is kept at a small value to generate chaotic dynamics in QLRNNM. $r_{l,l}$ is also small so that chaotic dynamics could be introduced in M-neurons even though $r_{u,l}$ is 0. Presynaptic neurons receive rough target directional information from four directional microphones with a certain frequency and modulatory interneuron monitor obstacle information from ultrasonic sensors. Without obstacles preventing the robot moving forward, rough target directional information given to S-neurons results in adaptive dynamics in M-neurons, which enables the robot to approach the target. When there are obstacles in the detection range of ultrasonic sensor, modulatory interneuron become excited. This causes presynaptic inhibition and weakens the response sensitivity of M-neurons to S-neurons. Correspondingly, complex dynamics emerging in motor neurons could make the robot find an appropriate detour to avoid obstacles.

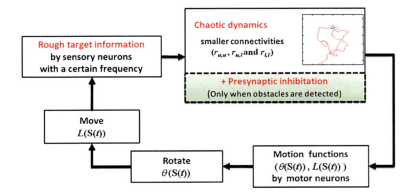

Fig. 9.14 Control algorithm for solving 2-dimensional mazes

Fig. 9.15 Examples of computer experiment: (**a**) Without obstacle, the robot can easily reach the target only depending on rough directional information; (**b**) When there are obstacles preventing the robot reaching the target, the robot finds a detour to avoid them and finally reaches the target

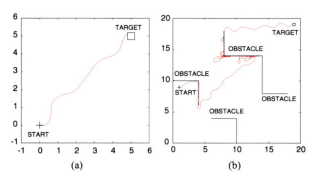

9.6 Computer Experiments and Hardware Implementation

Can the algorithm proposed above enable robots to solve two-dimensional maze tasks? In order to answer this question, several types of two-dimensional maze tasks have been demonstrated using computer experiments and actual hardware implementations.

The results of computer experiments indicate that, using the simple algorithm, chaotic dynamics in QLRNNM can be utilized to solve two-dimensional mazes successfully. Two examples are shown in Fig. 9.15. In the case of no obstacle to prevent the robot moving forward, the robot can efficiently move toward the target due to the sensitive response to sound input. On the other hand, when obstacles prevent it from moving forward, presynaptic inhibition ensures that the robot can find an appropriate detour to avoid them.

Now let us implement chaotic dynamics in an actual hardware implementation which solves complex control tasks. The hardware implementation was done in two steps. First, no obstacles are set between the robot and the target so that we can check the direct response of chaos. An example of the experiment without obstacles is shown in Fig. 9.16. The robot approaches the target along a trajectory similar to that in Fig. 9.15(a). Second, obstacles are set between the robot and the target so that we can check the effect of presynaptic inhibition. Figure 9.17(a) shows the experiment configuration where a loud speaker is set as a target and obstacles consist of walls forming a typical 2-dimensional maze. The results of the experiments with the hardware implementation also demonstrate that the robot can solve two-dimensional mazes using chaotic dynamics which emerge in both sensory and motor neurons. Some snapshots from a video of the experiment with the hardware implementation are shown in Fig. 9.17(a)∼(d).

Both computer experiments and actual hardware implementation show that the robot moves along a non-direct, zigzag path to reach the target. This raises some questions about efficiency. Generally speaking, efficiency is related with experience or cumulative learning. We think that adaptability and using experience are the two important elements of intelligence. In our present work, we pay attention to adaptability. The results of both computer experiments and experiments with actual hardware implementation, support our conjecture that complex dynamics emerging

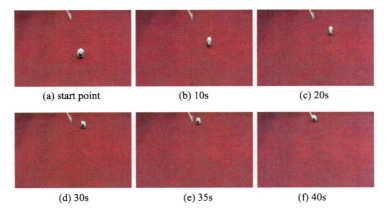

Fig. 9.16 Video snapshots of the robot approaching the target without obstacle. The robot is initially at the start point (**a**), and then responds to sound signal and moves toward the target (**b**)~(**e**), and finally reaches the target (**f**)

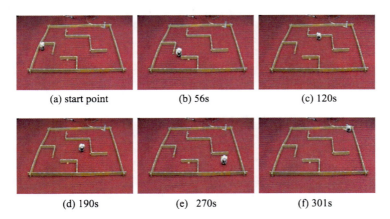

Fig. 9.17 Video snapshots of the robot solving a two-dimensional maze. The robot is initially at the start point (**a**), moves along a complex path, either toward the target or detouring to avoid obstacles (**b**), (**e**), and finally reaches the target (**f**)

in sensory-motor system could be used for adaptive dynamical behavior which is useful to solve complex problems with ill-posed properties.

9.7 Discussion

Is chaos really useful to control tasks? Towards the answer of this question, our work provides us with an indirect evidence for the potential possibility of chaotic dynamics in solving complex control tasks. Significantly, biological implication observed in our experiments should be noted. First, although we don't explicitly say that the crickets show a chaotic behavior for mating in maze areas, we have known that, from researches on auditory behavior of crickets, when one female cricket walks toward one singing male cricket, it always show a zigzag or meandering path for making suitable corrective turns [23]. Though we do not intend to simulate the cricket behavior, we can conclude that our experimental model could mimic auditory behavior of crickets to a certain extent. Second, recent works on brain-machine interface and the parietal lobe suggested that, in cortical areas, the "message" defining a given hand movement is widely disseminated [18, 24]. This means that there is a big reservoir of redundancy in brain. In our study, the whole state of neurons is utilized, so our approach also has a big reservoir of redundancy which results in great robustness observed in our experiments. Third, animals are often forced to deal with numerous ill-posed problems, but they exhibit amazing adaptive capability for doing them with easy. In our opinion, there could exist underlying simple rules to implement such excellent functions. In our work, chaotic dynamics have been applied successfully to solving complex problems in motion control via simple rules. This enables us to conclude that chaos could be an important role to maintain adaptability and creativity in biological systems including brain.

Based on these previous work, some problems are valuable to proceed in depth. In human motor control, there are more ill-posed control problems, such as adaptive gaits in changing environmental circumstances, simultaneous regulation of arms and legs, optimal trajectory selection, and so on. We believe that chaotic dynamics still have great potential that can be exploited for solving these complex problems.

9.8 Summary and Concluding Remarks

In this chapter, based on a viewpoint that chaotic dynamics could play an important role in complex information processing and control in biological systems, including brains, we have proposed a novel idea to harness the onset of complex nonlinear dynamics. Chaotic dynamics are introduced into QLRNNM and simple rules are used to apply the chaotic dynamics to complex control tasks such as solving two-dimensional mazes, which are ill-posed problems. An autonomous control system for a roving robot has been implemented. The robot is placed in a certain initial position, where the robot has no pre-knowledge about the position of target or obstacles, and it has to find the target fully autonomously only depending only on rough target direction sensing using sound signals. Both computer experiments and hardware implementation show that chaotic dynamics has the potential to be used for solving complex control tasks with ill-posed properties. In closing this chapter, let us emphasize the following points.

- Chaotic dynamics were introduced into a quasi-layered recurrent neural network (QLRNNM) consisting of sensory neurons and motor neurons.
- QLRNNM was implemented in a roving robot that was designed to solve ill-posed problems (2-dimensional mazes).

- The robot can reach the target using chaotic dynamics which emerge in both sensory neurons (sensitive response to external input) and motor neurons (complex motion generation).
- When there is an obstacle, the robot can find an appropriate detour to avoid it using chaotic dynamics resulting from presynaptic inhibition.
- Complex dynamics emerging in a sensory-motor system can be exploited for adaptive dynamical behavior which is useful to solve complex problems with ill-posed properties.
- Chaotic dynamics have novel potential for complex control via simple rules.

Acknowledgements This work has been supported by Grant-in-Aid for Promotion of Science # 19500191 in Japan Society for the Promotion of Science.

References

1. Aertsen A, Erb M, Palm G (1994) Dynamics of functional coupling in the cerebral cortex: an attempt at a model based interpretation. Physica D 75:103–128
2. Amari S (1977) Neural theory of association and concept-formation. Biol Cybern 26:175–185
3. Freeman WJ (1987) Simulation of chaotic EEG patterns with a dynamic model of the olfactory system. Biol Cybern 56:139–150
4. Fujii H, Ito H, Aihara K, Ichinose N, Tsukada M (1996) Dynamical cell assembly hypothesis—theoretical possibility of spatio-temporal coding in the cortex. Neural Netw 9(8):1303–1350
5. Haken H (1988) Information and self-organization. Springer, Berlin
6. Haken H (1996) Principles of brain functioning. Springer, Berlin
7. Huber F, Thorson H (1985) Cricket auditory communication. Sci Am 253:60–68
8. Kaneko K, Tsuda I (2000) Complex systems: chaos and beyond. Springer, Berlin
9. Kay LM, Lancaster LR, Freeman WJ (1996) Reafference and attractors in the olfactory system during odor recognition. Int J Neural Syst 4:489–495
10. Li Y, Nara S (2008) Application of chaotic dynamics in a recurrent neural network to control: hardware implementation into a novel autonomous roving robot. Biol Cybern 99:185–196
11. Li Y, Nara S (2008) Novel tracking function of moving target using chaotic dynamics in a recurrent neural network model. Cogn Neurodyn 2:39–48
12. Mikami S, Nara S (2003) Dynamical responses of chaotic memory dynamics to weak input in a recurrent neural network model. Neural Comput Appl 11(3–4):129–136
13. Nara S (2003) Can potentially useful dynamics to solve complex problems emerge from constrained chaos and/or chaotic itinerancy? Chaos 13(3):1110–1121
14. Nara S, Davis P (1992) Chaotic wandering and search in a cycle memory neural network. Prog Theor Phys 88:845–855
15. Nara S, Davis P (1997) Learning feature constraints in a chaotic neural memory. Phys Rev E 55:826–830
16. Nara S, Davis P, Kawachi M, Totuji H (1993) Memory search using complex dynamics in a recurrent neural network model. Neural Netw 6:963–973
17. Nara S, Davis P, Kawachi M, Totuji H (1995) Chaotic memory dynamics in a recurrent neural network with cycle memories embedded by pseudo-inverse method. Int J Bifurc Chaos Appl Sci Eng 5:1205–1212
18. Nicolelis MAL (2001) Actions from thoughts. Nature 409:403–407
19. Suemitsu Y, Nara S (2003) A note on time delayed effect in a recurrent neural network model. Neural Comput Appl 11(3–4):137–143

20. Suemitsu Y, Nara S (2004) A solution for two-dimensional mazes with use of chaotic dynamics in a recurrent neural network model. Neural Comput 16(9):1943–1957
21. Tokuda I, Nagashima T, Aihara K (1997) Global bifurcation structure of chaotic neural networks and its application to traveling salesman problems. Neural Netw 10(9):1673–1690
22. Tsuda I (2001) Toward an interpretation of dynamic neural activity in terms of chaotic dynamical systems. Behav Brain Sci 24(5):793–847
23. Weber T, Thorson J (1988) Auditory behavior of the cricket. J Comp Physiol A 163:13–22
24. Wessberg J, Stambaugh C, Kralik J, Beck P, Laubach M, Chapin J, Kim J, Biggs S, Srinivasan M, Nicolelis M (2000) Real-time prediction of hand trajectory by ensembles of cortical neurons in primates. Nature 408:361–365

Chapter 10
Time Scale Analysis of Neuronal Ensemble Data Used to Feed Neural Network Models

N.A.P. Vasconcelos, W. Blanco, J. Faber, H.M. Gomes, T.M. Barros, and S. Ribeiro

Abstract Despite the many advances of neuroscience, the functioning of telencephalic neuronal ensembles during natural behavior remains elusive. The analysis of continuous data from large neuronal populations recorded over many hours across various behavioral states presents practical and theoretical challenges, including an ever-increasing demand for computational power. The use of neural network models to analyze such massive datasets is very promising. Conversely, large-scale neuronal recordings are expected to provide key empirical data able to constrain neural network models. Both applications pose the problem of defining a time scale of inter-

---

N.A.P. Vasconcelos · W. Blanco · S. Ribeiro
Brain Institute, Federal University of Rio Grande do Norte (UFRN), Natal, RN, 59078-450, Brazil

N.A.P. Vasconcelos · H.M. Gomes
Department of Systems and Computation, Federal University of Campina Grande (UFCG), Campina Grande, PB, 58249-900, Brazil

N.A.P. Vasconcelos · T.M. Barros
Edmond and Lily Safra International Institute of Neuroscience of Natal (ELS-IINN), Rua Professor Francisco Luciano de Oliveira 2460, Bairro Candelária, Natal, RN, Brazil

N.A.P. Vasconcelos
Faculdade Natalense para o Desenvolvimento do Rio Grande do Norte (FARN), Natal, RN 59014-540, Brazil

N.A.P. Vasconcelos
Faculdade de Natal, Natal, RN, 59064-740, Brazil

J. Faber
Fondation Nanosciences & CEA/LETI/CLINATEC, Grenoble, 38000, France

S. Ribeiro (✉)
Neuroscience Graduate Program, Federal University of Rio Grande do Norte (UFRN), Natal, RN, 59078-450, Brazil
e-mail: sidartaribeiro@neuro.ufrn.br

S. Ribeiro
Psychobiology Graduate Program, Federal University of Rio Grande do Norte (UFRN), Natal, RN, 59078-450, Brazil

est, expressed as a choice of temporal bin size. Since the temporal range of synaptic physiology is orders of magnitude smaller than the temporal range of behavior, any analysis of neuronal ensemble data should begin with a comprehensive screening of the time scale. Notwithstanding, this procedure is seldom employed, for it substantially increases the computational requirements of the task. In this chapter we describe methods, procedures and findings related to bin size screening of the activity of neocortical and hippocampal neuronal ensembles. The data were obtained from rats recorded across the sleep-wake cycle before, during and after the free exploration of novel objects. We demonstrate that the estimation of firing rate probability distributions across different behavioral states depends crucially on the temporal scale used for the analysis. Next, we use a correlation-based method to show that the identification of object-related neuronal activity patterns also depends on bin size. Finally, we show that, within an adequate range of bin sizes, individual objects can be successfully decoded from neuronal ensemble activity using neural network models running in a parallel computer grid. The results suggest that mnemonic reverberation peaks at different time scales for different objects, and underscore the need for the temporal screening of empirical data used to feed neural network models.

10.1 Introduction

Similarly to what happens in other areas of science [16, 24], contemporary research in neuroscience demands the analysis of ever growing data sets, thus requiring an unprecedented increase in computational power [30, 37, 53, 54]. In the past decades, neuroscientists have been increasingly focusing their attention on the large-scale recording of neuronal ensembles [19, 25, 31, 38, 39, 42, 44, 45, 60]. One of the leading techniques to investigate neuronal ensembles is the chronic implantation of microelectrode arrays, which has been successfully employed to follow the electrical activity of hundreds of neurons distributed across many specific brain areas, over extended periods of time, while animals perform a variety of spontaneous or learned behaviors [5, 8, 10, 19, 20, 25, 26, 31, 38, 39, 42, 44, 50, 51, 60]. This technique was initially aimed at the basic problems of neural coding [42, 43, 52], but it eventually became essential for the development of brain machine interfaces [25, 40, 41, 46].

Among its many promises, large-scale neuronal recordings are expected to provide key empirical data able to constrain neural network models [33, 35]. By the same token, the data mining of neuronal ensemble activity with neural network models has been very successful [28, 30, 53], and holds enormous potential for future applications. Typically, the inputs of the neural network come from the computation of some measurement of neuronal activation within a given time interval, and the goal is to predict some behavioral features of choice. Regardless of the direction chosen (from real data to networks or vice-versa), one problem that must be addressed is defining a time scale of interest, expressed as a choice of temporal bin size. This chapter presents three examples of the practical and theoretical difficulties related to time scale analysis of neuronal ensemble data comprising action potentials (spikes). The three examples are derived from a single dataset comprising various natural behaviors of rats.

First, we present a temporal analysis of the basic statistics of spiking activity across the three major behavioral states of mammals: waking, slow-wave sleep (SWS) and rapid-eye-movement sleep (REM). The results are relevant for future studies of pattern classification and modeling of neuronal ensemble responses, and supply relevant statistical information about the data profile of freely behaving animals. Second, we describe how bin size impacts a multi-neuron correlation method for the investigation of mnemonic reactivation related to novel object exploration. The method measures the similarity degree between two matrices, *target* and *template*, by calculating the Pearson's correlation coefficient between them [32, 50]. We show that the method strongly depends on bin size, and that the mnemonic reverberation for different objects peak at different time scales. Finally, we show how time scale affects the binary classification of complex objects by neural network models fed with neuronal ensemble data obtained from the primary sensory cortices and hippocampus. To cope with the elevated computational cost of this investigation, we describe in detail a high performance computational (HPC) approach, using a distributed computer grid [16]. The HPC technique was employed to parallel-process a large set of binary artificial neural network classifiers trained to decide between two classes of neuronal activation patterns: (i) patterns associated to the contact intervals with a given object; and (ii) patterns associated to the contact intervals with other objects. Receiver Operating Characteristic (ROC) analyses involving populations of these binary classifiers were used to make inferences about object coding in the brain areas investigated. Our results demonstrate that complex object representations can be detected in cortical and subcortical areas, provided that bin sizes are not overly large (>500 ms). Interestingly, this boundary has the same magnitude of reaction time [29, 47], and may represent a minimum temporal frame for animal behavior.

10.1.1 Biological Experimental Setup

The behavioral experiments that originated the data analyzed were performed on adult male Long-Evans rats raised in captivity, without any previous contact with the objects used in the experiments. Microelectrode arrays were surgically implanted in specific brain areas [26, 50, 51]: the regions studied were the primary visual cortex (V1, layer 5), the primary somatosensory cortex (S1, layer 5) and the hippocampus (HP, CA1 field). The data acquired for further processing consisted of spike time sequences and behavioral information.

Figure 10.1 illustrates the experimental design. Three major behavioral states were monitored (WK, SWS and REM). After the animal habituated to the environment (recording box), four novel objects were placed inside it, and the animal was followed afterwards as the sleep-wake cycle progressed. The recordings were therefore divided in three parts: pre-exploration, exploration and post-exploration.

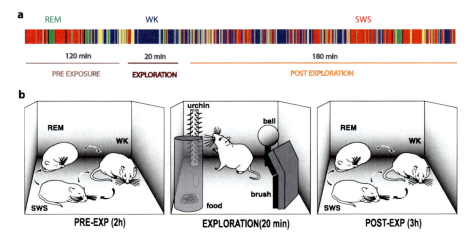

Fig. 10.1 Behavioral design. (**a**) Hypnogram, showing the sequence of behavioral states using a spectral analysis of local field potentials [17]; *yellow* indicates state transitions. (**b**) Experiments consisted of up to 6 hours of neuronal population and video recordings under infrared illumination. Four novel objects were introduced in the box for 20 minutes. They consisted of a golf ball attached to a spring ("ball"), a shoe brush mounted at an acute angle ("brush"), a tube with an opening filled with fruit loops ("food"), and a stick covered with outward thumbtacks ("urchin"). Drawing adapted from [51]

Throughout the entire experiment, behavioral and electrophysiological data were simultaneously acquired as previously described [44, 50, 51]. Behavioral states were automatically sorted by spectral analysis [17], producing a hypnogram as shown in Fig. 10.1a.

10.1.2 Neuronal Data Analysis

When properly stimulated, a neuron responds with a spike along its axon. It is possible to model the information processing carried out by the central nervous system in terms of spikes of groups of neurons, and sequences of spikes can be seen as the way the nervous system represents and communicates information across different brain areas [52]. When the main interest is the action potential, Eq. 10.1 presents a well-known model [7, 52] for a spike train, $\rho(t)$, where function $\delta(t)$ is the Dirac delta function, t_i is the i-th spike time within a spike train with S spikes:

$$\rho(t) = \sum_{i=1}^{S} \delta(t - t_i). \qquad (10.1)$$

Most algorithms designed to process spike trains involve the use of spike counts within fixed intervals (discrete bins) instead of the exact time when the spike occurred. Equation 10.2 defines the concept of firing rate [7], $r(t)$, for the interval $[t, t + \Delta t]$, considering the function of average response within that particular bin:

$$r(t) = \frac{1}{\Delta t} \int_{t}^{t+\Delta t} \langle \rho(\tau) \rangle d\tau. \quad (10.2)$$

Since the analysis of spike trains occurs for multi-neuron datasets, it is important to establish a way to represent a set of isolated neurons. Given $\boldsymbol{r}_j = [r_j(t_1) \ldots r_j(t_K)]$ the firing rate of the j-th recorded neuron, we have matrix **B**, which represents the electrical activation of a neuronal ensemble. The rows of $N \times K$ matrix **B** are formed with the firing rate values of the N neurons recorded along K bins:

$$\mathbf{B} = [\boldsymbol{r}_1 \quad \cdots \quad \boldsymbol{r}_N]^T. \quad (10.3)$$

10.2 Time-Dependent Statistics of Neuronal Firing Across the Major Behavioral States

It is well known that neuronal responses can be described statistically [7], and that the simplest way to carry out this task is to calculate the spike rate, which is obtained by counting the number of spikes that appear during a trial, and dividing this value by the duration of the trial. In order to extract the main statistical features of the dataset, in principle one could consider the entire experiment as a single block and determine the spike rate. However, with this procedure one would entirely miss the temporal perspective, that is, the variations in neuronal activity during the trial. To deal with this problem, one may define discrete bins that add up to the total recording time. This is a standard solution that captures the dynamic of neuronal activity over time, and can be defined as the time-dependent firing rate (Eq. 10.2). Bin size determination is a crucial step towards characterizing neuronal activity within a given state. For example, the use of small bin sizes (Δt) allows high temporal resolution but may greatly distort the biological phenomenon, because the majority of the spike counts on any given trial will be 0.

A simple approach to deal with the bin size problem is to represent the probability density function (PDF) of the time-dependent firing rate. Since extracellular recordings in freely behaving animals typically come from different brain regions and sleep-wake states, it is necessary to obtain the PDF separately for each variable of interest. Figure 10.2 represents the PDF of the time-dependent firing rate for hippocampus (HP) and the primary somatosensory cortex (S1). The PDFs obtained for a given animal, brain region or behavioral state are very different depending on the bin size selected. Figure 10.3 shows that the PDF means are constant within each state for bin sizes between 100 ms and 1000 ms. PDF variances, on the other hand, decrease substantially as bin size increases, and stabilize beyond 500 ms. Figures 10.2 and 10.3 complement each other, but a simple visual inspection may not reveal that the PDFs are different. This is the case for GE9, an animal with low and similar firing rates for all states. When dealing with cases like that, it is necessary to use another statistical metric to compare PDFs.

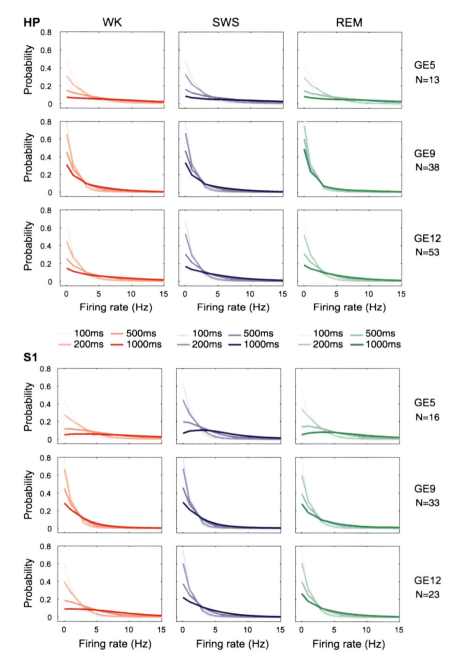

Fig. 10.2 Probability density functions of the time-dependent firing rate of neurons recorded from the HP (*top panels*) or S1 (*bottom panels*). Every row represents a different animal, where "*N*" indicates number of neurons used in that row

Statistical comparisons using the Kolmogorov–Smirnov test [34, 36, 56] are shown in Fig. 10.4, comparing states for 4 specific bin sizes. These results show that the majority of the PDFs are not similar. One exception was animal GE12, with statistically identical PDFs for SWS and REM, for bin sizes equal to 100 ms and 200 ms. However, for an adequate choice of bin size between 500 ms and 1000 ms, the PDFs of SWS and REM were statistically distinguishable.

The Poisson distribution provides a useful approximation of stochastic neuronal firing [7]. To determine whether state-specific PDFs of neuronal activity can be modeled by the Poisson analytical distribution, the PDFs shown in Fig. 10.2 were compared with a homogeneous Poisson distribution, for which the firing rate is constant over time. For all the comparisons we obtained $h = 0$, meaning that the empirical distributions were not equivalent to Poisson distributions. Taken together, the results demonstrate that the basic statistics of neuronal firing rates are different across behavioral states, do not correspond to a Poisson distribution, and depend on bin size.

10.3 Temporal Dependence of Neuronal Ensemble Correlations Used to Measure Mnemonic Reverberation

A well-established method to investigate mnemonic reverberation in neuronal ensembles is the use of pairwise [49] or multi-neuron [32, 50, 57] correlations of firing rates. The multi-neuron approach measures the similarity degree between two matrices, *target* and *template*, by calculating the correlation coefficient between them [32, 50, 57]. In this section, we describe how bin size impacts on the multi-neuron template matching method for the investigation of mnemonic reverberation related to novel object exploration. As input data, we used simultaneously recorded spike trains of N neurons binned into T intervals of a specific size.

The target matrix was constructed by selecting SWS or REM episodes from the total hypnogram, and then concatenating all the episodes into a single block, either before (PRE) or after (POST) the exploration of novel objects. This procedure allowed the construction of behaviorally homogeneous target matrices. To guarantee that the data concatenation would not destroy important temporal features of the neuronal record, we applied the same procedure to the correlation time series calculated for each entire recording. By statistically comparing the '*a priori*' concatenation and the '*a posteriori*' concatenation of the correlation time series, we verified that the artificial boundaries between concatenated sleep periods did not produce information loss.

Templates of different bin sizes were chosen by selecting epochs of contact between the facial whiskers and each of the objects, based on the visual inspection of the videotaped behavior. To calculate the correlation time series, the templates were slid along the whole target matrix with a step of 15 bins. This step size was cho-

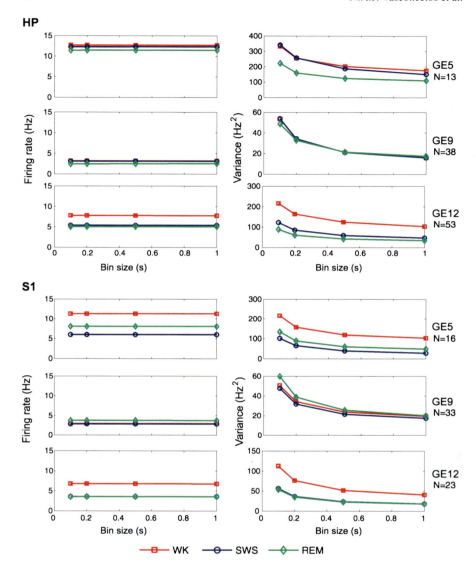

Fig. 10.3 Bin size dependence of means and variances of the probability density functions. Neurons recorded from HP (*top panels*) and S1 (*bottom panels*)

sen to promote fast computation with a minimum overlap in time for the smallest templates. For a $N \times T_{\text{tp}}$ template data matrix X and a $N \times T_{\text{tg}}$ target data matrix Y,

$$X = \begin{pmatrix} x_{11} & \cdots & x_{1T_{\text{tp}}} \\ \vdots & \ddots & \vdots \\ x_{N1} & \cdots & x_{NT_{\text{tp}}} \end{pmatrix}, \quad Y = \begin{pmatrix} y_{11} & \cdots & y_{1T_{\text{tg}}} \\ \vdots & \ddots & \vdots \\ y_{N1} & \cdots & y_{NT_{\text{tg}}} \end{pmatrix},$$

10 Time Scale Analysis of Neuronal Ensemble Data

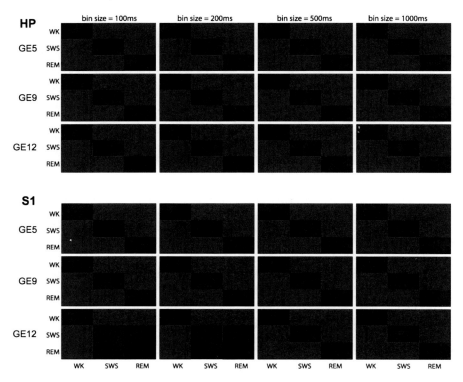

Fig. 10.4 Color-coded p values of the Kolmogorov–Smirnov (KS) test for neurons recorded from HP or S1. The PDFs shown in Fig. 10.2 were compared across behavioral states for different bin sizes. As expected, the main diagonal in each comparison has $p = 1$ (*red*), which means identity. A $p = 0$ (*blue*) means that the PDFs do not belong to the same continuous distribution, with an $\alpha = 0.05$

where N is the number of neurons, T_{tp} is the template size and T_{tg} is the target size, with $T_{\text{tg}} = T_{\text{tp}}$. The correlation time series $C(t)$ was obtained by:

$$C(t) = \frac{\sum_{i=1}^{N}\sum_{j=1}^{T_{\text{tp}}}(x_{ij}-\bar{x})(y_{ij}-\bar{y})}{\sqrt{\sum_{i=1}^{N}\sum_{j=1}^{T_{\text{tp}}}(x_{ij}-\bar{x})^2}\sqrt{\sum_{i=1}^{N}\sum_{j=1}^{T_{\text{tp}}}(y_{ij}-\bar{y})^2}}, \quad (10.4)$$

in which we take $T_{\text{tg}} \equiv T_{\text{tp}}$ at each time step t and the means $\bar{x}$ and $\bar{y}$ are defined by

$$\bar{x} = \frac{1}{NT_{\text{tp}}}\sum_{i=1}^{N}\sum_{j=1}^{T_{\text{tp}}}x_{ij}, \qquad \bar{y} = \frac{1}{NT_{\text{tp}}}\sum_{i=1}^{n}\sum_{j=1}^{T_{\text{tp}}}y_{ij}.$$

By definition, $C(t)$ is the 2-D Pearson's coefficient, with values ranging between 1 and -1. Since the method involves bi-directional averages along rows and columns of the template and target matrices, we normalized each row (neuron) by the maximum firing rate found in any bin, to account for possible miscorrelation effects due to absolute firing rate values. We also conducted a subtraction of temporal correlations derived from surrogated templates, as follows:

$$\rho_n(t) = \left\| C_{\text{novelty}}(t) - C_{\text{surrogate novelty}}(t) \right\|, \tag{10.5}$$

where $\|\cdot\|$ is the Euclidean norm. The surrogated templates were calculated by randomly shuffling rows and columns. This surrogate method destroys temporal and spatial information but preserves firing rates, and therefore contributes to the minimization of spurious effects caused by rate fluctuations.

After calculation of the correlation time series, the results were averaged according to object identity ($n = 4$ templates/object). Figure 10.5 shows that the mnemonic reverberation of each object during SWS is very sensitive to the temporal scale chosen for the investigation, because each object reverberates maximally at different bin sizes. Furthermore, a major increase in the correlations calculated for the post-novelty block, which has been proposed to be a hallmark of mnemonic reverberation [11, 12, 57], does not occur for all bin sizes. Compare, for instance, the correlations calculated for two different objects in GE5: a post-novelty increase in the correlations is clearly observed for "brush" using a 250 ms bin size, but for "ball" the effect is only noticeable at a bin size of 500 ms.

In summary, the data suggest that complex objects are encoded at distinct time scales. Such temporal multiplexing of information is likely related to the various spatial frequencies of the stimuli, as well as to the variations in the exploratory movements performed by the animal. The results demonstrate that one cannot integrate multi-neuron correlations across animals or brain regions using the same set of temporal parameters.

10.4 Object Classification by Neural Networks Fed with Ensemble Data: Bin Size Screening, Neuron Dropping and the Need for Grid Computation

Artificial neural networks are very useful for the analysis of neuronal activation recordings with the aim of predicting or explaining behavior [23, 30, 53, 59]. In our experiments, we used neural networks as binary classifiers for the identification, exclusively from the neuronal ensemble data, of the four different objects used in the recordings. As described in Sections 10.2 and 10.3, the statistics of neuronal ensemble activity are strongly dependent on bin size. Therefore, it is wise to begin any neural network analysis of neuronal ensemble data with a comprehensive investigation of time scales [52]. Another common question one faces when analyzing the activity of neuronal populations is the dependence on ensemble size, usually addressed by the neuron dropping method [30, 59]. However, both bin size screening and neuron dropping analysis are computationally expensive. Grid computing represents a recent solution for the problem of providing high-performance computing infrastructure. The denomination *grid computing* is used when the resource usage is symmetrical between users and computers, and when there is a continuous exchange

10 Time Scale Analysis of Neuronal Ensemble Data

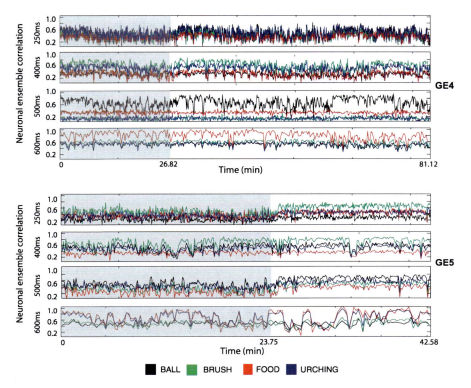

Fig. 10.5 Neuronal ensemble correlations depend strongly on bin size. Correlations were measured over the entire SWS record using a sliding window to match templates of neuronal ensemble activity sampled from episodes of whisker contact with one of four objects (ball, brush, food, urchin). *Grey segments* represent time prior to object exploration, *white segments* represent time after object exploration. Data correspond to neuronal ensembles recorded from the hippocampus

of computing resources among participants. These computer resources are mostly suitable for *Bag of Task* (BoT) applications, that is, which can be decomposed as a set of independent tasks. Grid computation is much cheaper (in terms of equipment and maintenance) than traditional supercomputing and cluster approaches. Differently from traditional high-performance computing, grids tend to be heterogeneous and geographically dispersed. This concept inspired on the Internet seeks to harness the growing number of computers connected to the World Wide Web, by designing strategies that use computer networks for high performance distributed computation [3, 14–16]. Grid strategies have been very successful in recent years, reaching a computing power of 100 teraflops [2, 6].

Grids are often built in terms of general-purpose software libraries known as middleware. Among the several grid middlewares available, we have chosen the

Ourgrid[1] because it is simple, fast and scalable [6]. Other alternatives are the BOINC (Berkeley Open Infrastructure for Network Computing) [1, 2] and Globus [14–16].

In the grid computing terminology, an application is termed as *job*, $J_i = \{t_{i,1}, \ldots, t_{i,L}\}$, which is defined by a set of *tasks*. The duration of a task, $t_{i,j}$, is defined by Eq. 10.6, where, $begin(t_{i,j})$ and $end(t_{i,j})$ correspond to the start and end time of completion, respectively, of a successful run of task $t_{i,j}$.

$$d(t_{i,j}) = end(t_{i,j}) - begin(t_{i,j}). \tag{10.6}$$

The total completion time, *makespan*, m_i, of a given job, $J_i = \{t_{i,1}, \ldots, t_{i,L}\}$, is given by Eq. 10.7.

$$m_i = \max\{end(t_{i,j})\} - \min\{begin(t_{i,j})\}, \quad t_{i,j} \in J_i. \tag{10.7}$$

For a given job, $J_i = \{t_{i,1}, \ldots, t_{i,L}\}$, the *speedup*, s_i, for a parallel execution of the job, is given by Eq. 10.8.

$$s_i = \frac{\sum_{i=1}^{L} d(t_{i,j})}{m_i}. \tag{10.8}$$

The job speedup, defined in Eq. 10.8, may not be adequate to measure performance improvement after parallel execution, so we propose another performance measure: *speedup in interval*. Given a time interval, $I_k = [b_k; e_k]$, it is said that a task, $t_{i,j}$, happens in I_k when, $end(t_{i,j}) \leq e_k$ and, $begin(t_{i,j}) \geq b_k$. Here, this relation is denoted by $t_{i,j} \sqsubset I_k$, and the length of the interval is given by $|I_k| = e_k - b_k$. The measure of *speedup in interval*, I_k, is defined by Eq. 10.9:

$$\bar{s}_k = \frac{\sum_{t_{i,j} \sqsubset I_k}^{L} d(t_{i,j})}{m_i}. \tag{10.9}$$

10.4.1 Object Classification

In this chapter, we present results obtained with the Multi-Layer Perceptron (MLP), a well-established neural network model [18] reliably used for input classification in neuroscience [30, 59]. In the MLP, artificial neurons are organized in three kinds of layers: *input* (simply for receiving the input signals); *hidden* (one or more processing layers bridging the input to the output layer); and *output* (to compute the network output values) [18].

Neural network weights are typically computed using a learning algorithm that iteratively approximates input to output mappings. The algorithm uses one sample set for training the association between a given input and the desired output, and another sample set for validation of the results, so as to prevent data over-fitting [27]. For this chapter, we employed the Powell–Beale variation of the conjugate gradient

---

[1] http://www.ourgrid.org.

learning algorithm [4, 48], because it presents a much better convergence performance than the traditional methods [9]. After being trained, a neural network can be evaluated using datasets with known output values but not used for training. By comparing the actual network outputs with the known outputs it is possible to calculate error rates related to false positives (acceptance) or false negatives (rejection).

We evaluated the quality of a population of binary classifiers along a space of contexts formed by the Cartesian product of the following parameters: animal (6 animals in total); objects (ball, brush, food and urchin), brain area (group of all recorded neurons, ALL, HP, S1 and V1); and, a parameter (SEP—Special Evaluation Parameter) that can be the size of the bin (*bin_size*), or the amount of neurons in each monitored brain area (*num_neurons*). For each context $c \in \Lambda$, it was necessary to evaluate a population of size L, of binary classifiers, $\Omega_{i,c}$. By plotting a graph of false acceptance versus false rejection (Receiver Operating Characteristic—ROC curve) [13], it is possible to assess the overall performance of the neural network used as a classifier. The quality Q of each binary classifier, $Q(\Omega_{i,c})$, corresponds to the *area* under the ROC curve of the binary classifier, $\Omega_{i,c}$. Positive samples were defined as the neuronal ensemble activity patterns sampled at the beginning of the animal's contact with the target object, and the negative samples were obtained by choosing randomly from the activation patterns observed during the animal's contact with other objects. For a context $c \in \Lambda$, the vector $\boldsymbol{Q}(c)$ is defined by Eq. 10.10, which encodes the quality of each one of the evaluated binary classifiers in that context:

$$\boldsymbol{Q}(c) = \big[Q(\Omega_{1,c}), \ldots, Q(\Omega_{L,c}) \big]. \tag{10.10}$$

The individual classifier evaluations were idempotent, therefore characterizing a BoT problem, which fits well in the target class of problems of the OurGrid middleware [6]. The input to the MLP were time windows $\mathbf{W}_k$ of the binned matrix defined in Eq. 10.3 (Fig. 10.6), obtained after a normalization operation, $f(\cdot)$.

Finally, we evaluated object coding using a population of classifiers ($n = 10$). In order to minimize any biases, each classifier was created using a different initialization. Furthermore, the training and testing sets were disjoint from each other. This method allows for experimental replication, and is more precise than the simple analysis of a few representative classification tests. Figure 10.7 shows that it is possible to reach a classification quality greater than 0.75 of average AUROC for 3 of the 4 objects.

To assess the temporal scale of object classification by neuronal ensembles, we evaluated the interval used in Eq. 10.2 to build the binned matrix **B** (defined in Eq. 10.3). The parameter SEP has been defined as the bin size used to setup the input matrices for the classifiers, $\Omega_{i,c}$, for each context, $c \in \Lambda_{bs}$. This parameter also plays a role in the evaluation of classification quality for each context. Figure 10.8a shows the impact of time scale on the classification quality of the "ball" object in two different animals (GE6 and GE13). The results were obtained for 15 different bins sizes.

Considering the results shown in Fig. 10.8 and the specific freely behaving protocol used in our experiment, in which for some objects there are very few contact

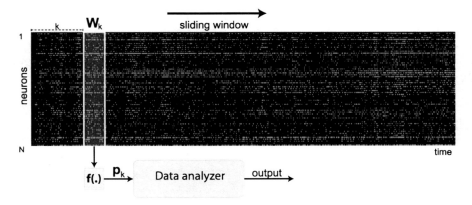

Fig. 10.6 Overall data processing scheme of a binned matrix of neuronal spikes. Data segments from episodes of object exploration are taken using a sliding window of size 2.5 s and position k. The data are fed as input to a binary classifier, whose output represents a "yes" or "no" to the question of whether a given object was explored during a given data segment

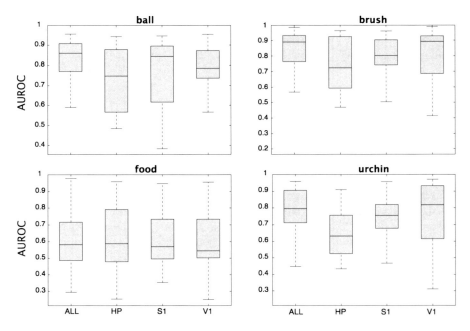

Fig. 10.7 General view of the classification results. The *boxplots* show medians and quartiles. "*All*" corresponds to results obtained from a pool of all neurons recorded in the three different brain areas. High AUROC values, indicative of correct object classification, were observed for three of the four objects [58]

samples, we concluded that 250 ms was the best bin size to perform object classification based on neuronal ensemble activation as input data fed to a MLP. As

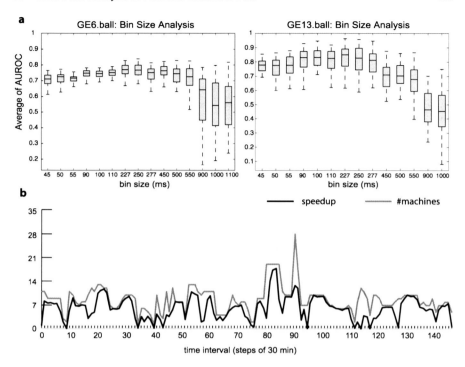

Fig. 10.8 Bin size analysis results. (**a**) Box plots of AUROC results. Bin sizes within 45–500 ms yielded distributions of average AUROCs with a median of approximately 0.7. For bin sizes above 550 ms the median of the distributions of average AUROCs decayed, decreasing to 0.5 at bin sizes around 1000 ms. (**b**) Temporal measures of speedup and number of participant machines in the computer grid applied to the bin size comparison. This experiment included two sites linked by a 512 Kbs VPN tunnel

can be seen in the Fig. 10.8b, there is good agreement between speedup and the corresponding number of machines added to the grid. For the particular time interval shown in Fig. 10.8b, we used a relatively small number of machines in the grid (61 machines). Only 3 of these machines were fully dedicated to the job; the other machines were desktops which were assigned to the grid during their times of idleness. The average speedup was around 7, and the average number of machines was 10. Even with this small speed up and the small number of machines, when compared with larger grid solutions [14–16], the results provide a successful example of increased computational power based on a simple, fast and scalable [6] grid framework.

10.4.2 Neuron Dropping Analysis

Neuron dropping is a method used to investigate the relationship between classification quality, $Q(c)$, for a given context, $c \in \Lambda_{nd}$, and the amount of neurons used for the classification [30, 41]. Figure 10.9 contains a representative result for the neuron dropping analysis. Given a number of neurons, n, the average classification quality is calculated based on the evaluation of $K = 10$ different classifiers, each one trained and tested with a dataset generated based on the activation information of n neurons randomly chosen. This information (number of neurons *versus* average classification quality) is represented by dots of different colors for different neuroanatomical areas (HP, S1 and V1). The results in Fig. 10.9a indicate that neuronal sets of around 10 neurons allow for good object classification, comparable to the results obtained with larger neuronal ensembles (>30 neurons). This computation took approximately 4 days on the Ourgrid platform. Figure 10.9b depicts speedup graphs for the neuron dropping analysis, as defined in Eq. 10.5, as well as the number of machines used over time. There was good correspondence between speedup and the number of machines included in the grid. At some point there were 89 machines participating in the grid; however, only 3 of these machines were fully dedicated to the analysis, the other computers were desktops which were assigned to the grid during their times of idleness. The average speedup was around 11, and the average number of machines was 13.

Higher speedup values were obtained in the neuron dropping analysis when compared to the speedup of the bin size analysis (Fig. 10.8). This can explained by more idleness and a higher number of machines in the grid used for the neuron dropping analysis (28 additional processors), as can be confirmed by the time intervals between 80 to 110 (Fig. 10.9b).

The context space, Λ_{nd}, results from the Cartesian product formed by characteristics which define the context: animal (4), object (4), area (4), and binsize (15). For each context, 50 classifiers were evaluated. This determines a universe of 48,000 classifiers evaluated for the bin size experiment. Such a large set of classifier evaluations would take more than 87 days in sequential execution; but using the grid architecture described here, it took less than 13 days. For the neuron dropping analysis, the context also included the number of neurons (<10), leading to a total of approximately 100,000 classifiers evaluated for the neuron dropping analysis. This evaluation would take more than 85 days in a sequential execution; but using the grid architecture the computation took less than 12 days.

10.5 Conclusion

In this chapter, we demonstrated that the use of neuronal ensemble data to feed neural networks is highly dependent on the temporal scale chosen for the analysis. We showed that a cooperative grid architecture can efficiently implement computationally expensive analyses based on neural networks, such as neuron dropping and bin size screening. The means and variances of neuronal firing rates recorded from the primary sensory cortex and the hippocampus across the sleep-wake cycle vary according to bin size, but stabilize beyond 500 ms. A multi-neuron correlation measure of mnemonic reverberation also varies according to bin sizes. Different objects seem to be encoded at distinct time scales, revealing a temporal multiplexing

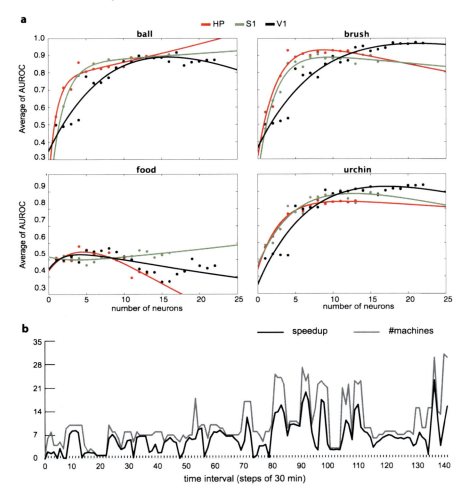

Fig. 10.9 Neuron dropping results. (**a**) Neuron dropping curves described the fit of a double exponential for HP, S1 and V1 (*green, red* and *black*, respectively) for the different objects. Typically, classification begins with low AUROC values (around 0.5) for small numbers of neurons, and increases until the number of neurons is around $n = 10$, stabilizing afterwards. Similar curves were observed in all anatomical areas studied. (**b**) Temporal measures of speedup and number of participant machines in the computer grid applied to the neuron dropping task. This experiment included two sites linked by a 512 Kbps VPN tunnel

of information that is likely related to the various spatial frequencies of the stimuli, as well as to the variations in the exploratory movements performed by the animal. An increase in correlations after novel experience, one of the tenets of mnemonic reverberation [11, 12, 57], occurs at different bin sizes for different animals (250–500 ms). The neuron dropping analysis provided evidence that complex object coding in the neocortex and the hippocampus occurs at the level of distributed neuronal ensembles. We also showed that complex object classification based on neuronal ensemble data depends on bin size. The bin size screening demonstrated that there exists a wide range of temporal scales up to ~500 ms that allow for accurate object classification. Notably, this upper boundary is of the same order of magnitude of reaction time [29, 47] and may actually represent a minimum temporal frame for overt behavior. These times are at least two orders of magnitude larger than those thought to be relevant for mnemonic encoding at the neuronal ensemble level [21, 22, 32, 55]. Future studies shall elucidate the relationship between these two levels of description of the behaving brain. Overall, the results underscore the need to investigate the temporal dimension when using neural data from freely behaving animals to feed neural networks.

Acknowledgements Work on this chapter was supported by Coordenação de Aperfeiçoamento de Pessoal de Nível Superior (CAPES), Conselho Nacional de Desenvolvimento Científico e Tecnológico (CNPq) and Associação Alberto Santos Dumont para Apoio à Pesquisa (AASDAP). We thank Adriana Ragoni and Marcelo Pacheco for laboratory management; Lineu Silva at Faculdade Natalense para o Desenvolvimento do Rio Grande do Norte (FARN), Ruphius Germano at Faculdade de Natal (FAL) and Diogo Melo (Forte Network) for technical support to set up the metropolitan computer grid in Natal. We also thank the Ourgrid Team for technical support on Ourgrid middleware.

References

1. Anderson DP (2004) BOINC: a system for public-resource computing and storage. In: Grid computing, 2004. Proceedings. Fifth IEEE/ACM international workshop on, pp 4–10
2. Anderson DP, Cobb J, Korpela E, Lebofsky M, Werthimer D (2002) SETI@home: an experiment in public-resource computing. Commun ACM 45(11):56–61
3. Baker M, Buyya R, Laforenza D (2002) Grids and grid technologies for wide-area distributed computing. Softw Pract Exp 32(15):1437–1466
4. Beale EM (1972) A derivation of conjugate-gradients. In: Lootsma FA (ed) Numerical methods for nonlinear optimization. Academic Press, London
5. Chelazzi L, Miller EK, Duncan J, Desimone R (1993) A neural basis for visual search in inferior temporal cortex. Nature 363(6427):345–347
6. Cirne W, Brasileiro F, Andrade N, Costa L, Andrade A, Novaes R et al (2006) Labs of the world, unite!!!. J Grid Comput 4(3):225–246
7. Dayan P, Abbott LF (2001) Theoretical neuroscience: computational and mathematical modeling of neural systems. MIT Press, Cambridge
8. Deadwyler SA, Bunn T, Hampson RE (1996) Hippocampal ensemble activity during spatial delayed-nonmatch-to-sample performance in rats. J Neurosci 16(1):354–372
9. Demuth H, Beale M (2006) Neural network toolbox: for use with MATLAB. The Math Works Inc, Natick
10. Engel AK, König P, Kreiter AK, Singer W (1991) Interhemispheric synchronization of oscillatory neuronal responses in cat visual cortex. Science 252(5010):1177–1179
11. Faber J, Nicolelis MA, Ribeiro S (2008) Analysis of neuronal reverberation during slow-wave sleep after novelty exposition. In: I IBRO/LARC congress of neurosciences of Latin America, Caribbean and Iberian peninsula, Búzios
12. Faber J, Nicolelis MA, Ribeiro S (2009) Entropy methods to measure neuronal reverberation during post-learning sleep. In: XXXIV Society of Neuroscience meeting, Chicago
13. Fawcett T (2006) An introduction to ROC analysis. Pattern Recognit Lett 27(8):861–874

14. Foster I (2006) Service-oriented science: scaling the application and impact of eResearch. In: IEEE/WIC/ACM international conference on intelligent agent technology (IAT'06), pp 9–10
15. Foster I, Kesselman C (2004) The Grid 2: blueprint for a new computing infrastructure. Morgan Kaufman, San Mateo
16. Foster I, Kesselman C, Tuecke S (2001) The anatomy of the grid: enabling scalable virtual organizations. Int J High Perform Comput Appl 15(3):200–222
17. Gervasoni D, Lin S, Ribeiro S, Soares ES, Pantoja J, Nicolelis MA et al (2004) Global forebrain dynamics predict rat behavioral states and their transitions. J Neurosci 24(49):11137–11147
18. Haykin S (2008) Neural networks and learning machines, 3rd edn. Prentice Hall, New York
19. Hirase H, Leinekugel X, Czurkó A, Csicsvari J, Buzsáki G (2001) Firing rates of hippocampal neurons are preserved during subsequent sleep episodes and modified by novel awake experience. Proc Natl Acad Sci USA 98(16):9386–9390
20. Hoffman KL, McNaughton BL (2002) Coordinated reactivation of distributed memory traces in primate neocortex. Science 297(5589):2070–2073
21. Ikegaya Y, Aaron G, Cossart R, Aronov D, Lampl I, Ferster D et al (2004) Synfire chains and cortical songs: temporal modules of cortical activity. Science 304(5670):559–564
22. Ji D, Wilson MA (2007) Coordinated memory replay in the visual cortex and hippocampus during sleep. Nat Neurosci 10(1):100–107
23. Jianhua D, Xiaochun L, Shaomin Z, Huaijian Z, Yu Y, Qingbo W et al (2008) Analysis of neuronal ensembles encoding model in invasive brain-computer interface study using radial-basis-function networks. In: 2008 IEEE international conference on granular computing, pp 172–177
24. Juhász Z, Kacsuk P, Kranzlmüller D (2005) Distributed and parallel systems: cluster and grid computing. Springer, Berlin
25. Kim HK, Carmena JM, Biggs SJ, Hanson TL, Nicolelis MA, Srinivasan MA et al (2007) The muscle activation method: an approach to impedance control of brain-machine interfaces through a musculoskeletal model of the arm. IEEE Trans Biomed Eng 54(8):1520–1529
26. Kralik JD, Dimitrov DF, Krupa DJ, Katz DB, Cohen D, Nicolelis MA et al (2001) Techniques for long-term multisite neuronal ensemble recordings in behaving animals. Methods 25(2):121–150
27. Krogh A (2008) What are artificial neural networks? Nat Biotechnol 26(2):195–197
28. Krupa DJ, Wiest MC, Shuler MG, Laubach M, Nicolelis MA (2004) Layer-specific somatosensory cortical activation during active tactile discrimination. Science 304(5679):1989–1992
29. Kutas M, McCarthy G, Donchin E (1977) Augmenting mental chronometry: the P300 as a measure of stimulus evaluation time. Science 197(4305):792–795
30. Laubach M, Wessberg J, Nicolelis MA (2000) Cortical ensemble activity increasingly predicts behaviour outcomes during learning of a motor task. Nature 405:567–571
31. Logothetis NK, Pauls J, Poggio T (1995) Shape representation in the inferior temporal cortex of monkeys. Curr Biol 5(5):552–563
32. Louie K, Wilson MA (2001) Temporally structured replay of awake hippocampal ensemble activity during rapid eye movement sleep. Neuron 29:145–156
33. Markram H (2006) The blue brain project. Nat Rev, Neurosci 7(2):153–160
34. Marsaglia G, Tsang WW, Wang J (2003) Evaluating Kolmogorov's distribution. J Stat Softw 8(18)
35. Migliore M, Cannia C, Lytton WW, Markram H, Hines ML (2006) Parallel network simulations with NEURON. J Comput Neurosci 21(2):119–129
36. Miller LH (1956) Table of percentage points of Kolmogorov statistics. J Am Stat Assoc 51(273):111–121
37. Narayanan NS, Kimchi EY, Laubach M (2005) Redundancy and synergy of neuronal ensembles in motor cortex. J Neurosci 25(17):4207–4216
38. Nicolelis MA (2007) Methods for neural ensemble recordings (methods in life sciences—neuroscience section)

39. Nicolelis MA, Chapin JK (1994) Spatiotemporal structure of somatosensory responses of many-neuron ensembles in the rat ventral posterior medial nucleus of the thalamus. J Neurosci 14:3511–3532
40. Nicolelis MA, Chapin JK (2002) Controlling robots with the mind. Sci Am 287:46–53
41. Nicolelis MA, Lebedev MA (2009) Principles of neural ensemble physiology underlying the operation of brain-machine interfaces. Nat Rev, Neurosci 10(7):530–540
42. Nicolelis MA, Baccala LA, Lin RC, Chapin JK (1995) Sensorimotor encoding by synchronous neural ensemble activity at multiple levels of the somatosensory system. Science 268:1353–1358
43. Nicolelis MA, Ghazanfar AA, Faggin BM, Votaw S, Oliveira LM (1997) Reconstructing the engram: simultaneous, multisite, many single neuron recordings. Neuron 18:529–537
44. Nicolelis MA, Dimitrov D, Carmena JM, Crist R, Lehew G, Kralik JD et al (2003) Chronic, multisite, multielectrode recordings in macaque monkeys. Proc Natl Acad Sci USA 100:11041–11046
45. Nicolelis M, Ribeiro S (2002) Multielectrode recordings: the next steps. Curr Opin Neurobiol 12(5):602–606
46. Patil PG, Carmena JM, Nicolelis MA, Turner DA (2004) Ensemble recordings of human subcortical neurons as a source of motor control signals for a brain-machine interface. Neurosurgery 55
47. Posner MI (2005) Timing the brain: mental chronometry as a tool in neuroscience. PLoS Biol 3(2):e51
48. Powell MJ (1977) Restart procedures for the conjugate gradient method. Math Program 12:241–254
49. Qin YL, McNaughton BL, Skaggs WE, Barnes CA (1997) Memory reprocessing in corticocortical and hippocampocortical neuronal ensembles. Philos Trans R Soc Lond B, Biol Sci 352(1360):1525–1533
50. Ribeiro S, Gervasoni D, Soares ES, Zhou Y, Lin SC, Pantoja J et al (2004) Long-lasting novelty-induced neuronal reverberation during slow-wave sleep in multiple forebrain areas. PLoS Biol 2(1):e24
51. Ribeiro S, Shi X, Engelhard M, Zhou Y, Zhang H, Gervasoni D et al (2007) Novel experience induces persistent sleep-dependent plasticity in the cortex but not in the hippocampus. Front Neurosci 1(1):43–55
52. Rieke F, Warland D, Steveninck RD, Bialek W (1999) Spikes: exploring the neural code. MIT Press, Cambridge
53. Sanchez JC, Principe JC, Carmena JM, Lebedev MA, Nicolelis MA (2004) Simultaneus prediction of four kinematic variables for a brain-machine interface using a single recurrent neural network. In: Conference of the IEEE engineering in medicine and biology society, vol 7, pp 5321–5324
54. Schneidman E, Bialek W, Berry MJ (2003) Synergy, redundancy, and independence in population codes. J Neurosci 23:11539–11553
55. Siapas AG, Wilson MA (1998) Coordinated interactions between hippocampal ripples and cortical spindles during slow-wave sleep. Neuron 21(5):1123–1128
56. Stephens MA (1970) Use of the Kolmogorov–Smirnov, Cramer–Von Mises and related statistics without extensive tables. J R Stat Soc 32(1):115–122
57. Tatsuno M, Lipa P, Mcnaughton BL (2006) Methodological considerations on the use of template matching to study long-lasting memory trace replay. J Neurosci 26(42):10727–10742
58. Vasconcelos N, Pantoja J, Hindiael B, Caixeta FV, Faber J, Freire MAM, Cota VR et al (2011) Cross-modal responses in the primary visual cortex encode complex objects and correlate with tactile discrimination. Proc Natl Acad Sci USA, in press
59. Wessberg J, Stambaugh CR, Kralik JD, Beck PD, Laubach M, Chapin JK et al (2000) Realtime prediction of hand trajectory by ensembles of cortical neurons in primates. Nature 408:361–365
60. Wilson MA, McNaughton BL (1994) Reactivation of hippocampal ensemble memories during sleep. Science 265(5172):676–679

Chapter 11
Simultaneous EEG-fMRI: Integrating Spatial and Temporal Resolution

Marcio Junior Sturzbecher and Draulio Barros de Araujo

Abstract Electroencephalography (EEG) and functional Magnetic Resonance Imaging (fMRI) are among the most widespread neuroimaging techniques available to noninvasively characterize certain aspects of human brain function. The temporal resolution of EEG is excellent, managing to capture neural events in the order of milliseconds. On the other hand, its spatial resolution lacks precision. Conversely, fMRI offers high spatial resolution, typically on the order of mm^3. However, it has limited temporal resolution (~sec), which is determined by the haemodynamic response. Therefore, simultaneous acquisition of EEG and fMRI is highly desirable. In recent years, the ability to perform combined EEG-fMRI has attracted considerable attention from the neuroscience community. In this chapter, relevant methodological aspects of EEG-fMRI will be first described, focused on the nature of neurophysiologic signals and difficulties to integrate electric and haemodynamic signals derived from both techniques. Second, state of the art strategies related to artifact correction and signal analysis will be described. Finally, a possible use of EEG-fMRI will be presented, focused on its potential application in epilepsy.

11.1 Introduction

Studying the human brain has became one of the major challenges in science, and the integration of numerous observations related to cerebral activity can conceive a better understanding. Hemodynamic, metabolic, electric and magnetic signals are typical brain events among which a variety of neurophysiological processes can be derived. In humans, noninvasiveness is of the utmost importance. Therefore, in the early years, knowledge acquired on human brain function was actually inferred,

M.J. Sturzbecher · D.B. de Araujo
Department of Physics, FFCLRP, University of Sao Paulo, Ribeirao Preto, Brazil

D.B. de Araujo (✉)
Brain Institute, Federal University of Rio Grande do Norte, Natal, Brazil
e-mail: draulio@neuro.ufrn.br

D.B. de Araujo
Onofre Lopes University Hospital, Federal University of Rio Grande do Norte, Natal, Brazil

based on experiments conducted in small animals. Currently, however, the development of noninvasive techniques, also called functional neuroimaging, has changed this scenario.

One of the first functional neuroimaging techniques capable of examining some aspects of the human brain function was Positron Emission Tomography (PET). Improvements in the acquisition and processing radio frequency signals, and the development of fast acquisition protocols, have turned Magnetic Resonance Imaging (MRI) into a new alternative to inspect the human brain, through what it is called functional magnetic resonance imaging (fMRI). Moreover, the possibility of detecting electric and magnetic brain signals resulted in the appearance of electroencephalography (EEG) and magnetoencephalography (MEG).

The vast majority of such studies are based on the use of a single technique. On the other hand, in the last few years, there has been a great effort to accomplish an integrative, multimodal, approach. In particular, EEG and fMRI have been combined successfully to correlate the high temporal resolution of the bioelectric patterns detected by the EEG, with the high spatial resolution derived from fMRI.

Simultaneous EEG-fMRI acquisition can be integrated basically in asymmetric and symmetric fusion approaches [17]. Asymmetric integration is the use of information from one modality as a constraint to the other. In brief, either the temporal information from the EEG signal is used as a prior in the fMRI model [10, 68], or the spatial information from fMRI is used as predictor variable in the EEG inverse problem [60]. Unlike the asymmetric approach, which uses a preference signal as constraints, symmetrical data fusion uses models that link EEG and fMRI signals and extract features from both modalities [17].

Historically, simultaneous EEG-fMRI first aimed at epileptic patients [34]. However, new areas of interest such as pain [15] and anesthesiology research have been getting greater attention in the later years. Apart from clinical application, this methodology has been used to investigate spontaneous brain activity (delta, theta, alpha, and gamma rhythm) [20, 27, 51], the default mode network [44] or evoked potentials [12] in cognitive neurosciences.

In this chapter, relevant methodological aspects of EEG-fMRI will be first described, focused on the nature of neurophysiologic signals and difficulties to integrate electric and haemodynamic signals derived from both techniques. Second, state of the art strategies related to artifact correction and signal analysis will be described. Finally, a possible use of EEG-fMRI will be presented, focused on its potential application in epilepsy.

11.2 Origin of the EEG Signal

Richard Caton, in 1875, performed the first measurement of brain electrical activity by placing an electric field probe onto the surface of exposed brains of small animals. This study paved the way for future discovery of the alpha rhythm in the human EEG by Hans Berger in the early years of the 1920s [67]. Basically, neuronal signals detected by the scalp EEG are composed by extracellular electric currents that flow during synaptic excitations of the dendrites of many pyramidal neurons in gray matter. As a result of the high electrical resistivity of head layers (i.e., skull, scalp, cerebrospinal fluid and brain) and the distance of scalp electrodes to neurons, only a large number of neurons can generate recordable electric field potential on conventional EEG. Moreover, the magnitude of the field is highly dependent on the orientation of such ensemble of neurons, achieving a maximum when they are aligned perpendicularly to the brain surface.

The temporal resolution of EEG is excellent, managing to capture neural events in the order of milliseconds. On the other hand, its spatial resolution lacks precision. The location of where field potentials are generated implies the resolution of the inverse problem, that is, from the measured electric field one has to estimate the electric current distribution responsible for that field map. Unfortunately, this problem does not have a unique solution, which limits the precision of the reconstructed sources. This is also called an ill-posed inverse problem.

11.3 Origin of the fMRI Signal

In 1881, Angelo Mosso first observed that neuronal energy demand and cerebral blood flow are correlated (see [48], for review). Although the mechanisms that link neuronal activity to the brain vascular system are still subject of intense research, it is well known that neuronal activity causes increased consumption of ATP (adenosine triphosphate), which implies in an increased demand for glucose and oxygen. To meet the basic need for these substrates, there is a local increase in cerebral blood volume (CBV) and cerebral blood flow (CBF), which leads to an increased ratio of oxygenated hemoglobin (Oxy-Hb) with respect to deoxygenated hemoglobin (Deoxy-Hb). Furthermore, Pauling and Coryell discovered that the magnetic property of hemoglobin depends on whether or not it is bound to oxygen [62]. Oxy-Hb is diamagnetic, slightly decreasing the local static magnetic field, and Deoxy-Hb is paramagnetic, locally increasing the static magnetic field. As a net result, the decrease of Deoxy-Hb slightly enhances MR signal amplitude. This mechanism is known as BOLD (Blood Oxygenation Level Dependent).

Based on such property, Ogawa et al. [58] showed that deoxy-hemoglobin could act as an endogenous contrast agent in anesthetized rats when these animals were submitted to MRI evaluation. Soon after, it was demonstrated that the same mechanism could be applied to localize neuronal modulation in humans following visual and motor tasks [42, 59].

In human studies, fMRI offers high spatial resolution, typically on the order of mm^3. However, it has limited temporal resolution ($\sim$sec), which is determined by the relatively slow hemodynamic response.

11.4 Relationship Between EEG and fMRI

When functional neuroimaging modalities are to be combined it is essential that all signals relates to the same basic physiological phenomenon. It is well known that EEG is correlated with Local Field Potencial (LFP) in human brain. Although BOLD-fMRI has been used by more than a decade, only recently the correlation between neuronal activity and BOLD effect was demonstrated experimentally, by mapping the primary visual area (V1) of anesthetized and awake monkeys using simultaneous electrophysiology and fMRI in [26, 49]. This and further studies have shown a relatively linear relationship between haemodynamic responses, spiking activity and LFP in different brain areas of humans and non-humans (rats, cats, monkey) [33, 54, 70]. Although both Multi Unit Activity (MUA) and LFP contribute significantly to the BOLD signal, the later shows an additional sustained response that is maintained during the entire stimulus presentation. Other authors have also reported similar results using a variety of methods and mapping different brain areas [63, 74].

11.5 Acquisition of Simultaneous EEG-fMRI: Safety and Methodological Issues

Performing simultaneous EEG-fMRI has many technical and safety challenges that must be considered and overcome before its implementation. Indeed, artifacts (either in the MRI or on the EEG traces) and tissue damage can arise as a result of the electric nature of EEG signals.

A first relevant issue relates to the amount of radiofrequency energy deposited onto the electrodes wires and cable loops during the application of magnetic fields gradients, necessary for MRI generation. According to Faraday's law, any temporal change in magnetic flux can induce electric currents in electrodes and electrode leads [46], which can flow through the tissue and cause neural stimulation and/or heating. Therefore, only non-ferromagnetic electrodes and leads should be used: typically, leads are made of carbon fiber and electrodes of Ag/AgCl or gold. Moreover, all electrode plates must have current limiting resistance with a resistivity between 5 and 15 kOhms [34, 46].

Second, one needs to reduce the amount of artifact induced in the EEG traces. Basically, such artifacts can result from three different sources of interference: movement of conductive material, MR magnetic gradients and pulse artifacts.

Any movement of the electrodes or change of the magnetic field around the wires will induce electric currents that will contaminate and eventually obscure the EEG signal of interest. Movement artifacts can occur as a result of subject head motion, physiological motion (vascular and respiratory) or be related to movement of parts of the MRI scanner, such as the cryogen pump and gradient switching system [9]. Gradient artifacts (GA) appear, again, as a result of Faraday's law, where electromotive force is induced in the electrodes as a consequence of temporal changes in the magnetic fields values applied during MR imaging. GA is significantly larger in amplitude (up to several mV) than the physiological EEG and induces high frequency components to the traces [4].

11 Simultaneous EEG-fMRI: Integrating Spatial and Temporal Resolution

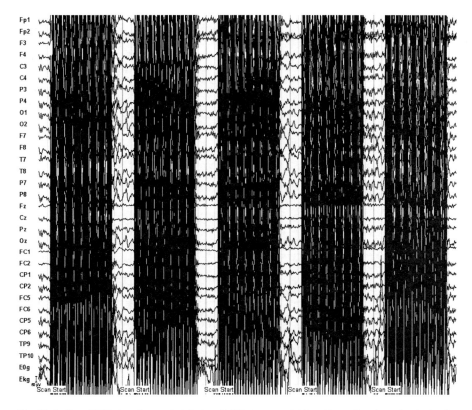

Fig. 11.1 Raw EEG signal recorded during fMRI data acquisition. '*Scan Start*' marks the beginning of each fMRI acquisition

Even with an optimized EEG-fMRI setup, artifacts still occur. Figure 11.1 illustrates a typical example of continuous EEG recorded inside the scanner during an fMRI session. Note the evident contamination by GA in the EEG traces.

Without computational signal processing, it would not be possible to obtain meaningful EEG recordings. The standard correction approach is based on an average artifact subtraction (AAS) algorithm [3, 4, 9, 31, 57, 76].

The basic principle of the AAS method is to assume similarity, that is, the repetitive nature of the artifact, and the absence of its correlation with physiological signals of interest. Each epoch can be identified looking for the onset of the slice or volume acquisition (Marked as "Scan Start" in Fig. 11.1). Averaging epochs allows one to create artifact templates for each channel, which can be subtracted from the data and thus recover the physiological signal of interest. To reduce residual image acquisition artifacts, an adaptive noise cancellation is applied [4]. Figure 11.2 shows a representative EEG interval after artifact reduction using the AAS method.

Another important piece of artifact comes in the form of periodic low frequency signals, which contaminate most EEG channels, the so-called pulse artifact, or ballistocardiogram (BCG). The exact origin of the BCG is still unknown but it is likely

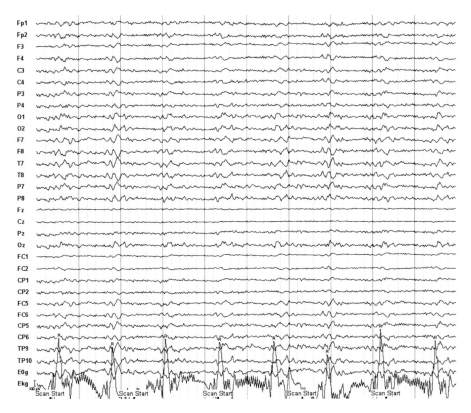

Fig. 11.2 The same EEG trace shown in Fig. 11.1 after correction of gradient artifacts using the average artifact subtraction (AAS) algorithm

caused by expansions and contraction of scalp vessels, associated with the cardiac cycle [3, 31, 34]. Pulsatile blood vessels move the EEG electrodes (or cables), leading to significant current induction (up to 200 μV at 3 Tesla). Figure 11.3 depicts an example of a periodic pulse artifact.

Unlike GA, removing BCG artifact is slightly more difficult since its periodicity is not completely regular, and is associated with heart rate variability. In addition, BCG amplitude and morphology varies between EEG channels and can differ across heartbeats. Even though a variety of methods have been developed to reduce these artifacts [3, 9, 72], AAS is again a relatively efficient alternative. Figure 11.4 shows the results after GA and PA correction using AAS.

The use of AAS approach to promote pulse artifact correction relies on the detection of each cardiac cycle onset in de Electrocardiogram (ECG) channel. As a result, an accurate template is achieved averaging all epochs. Although AAS has been successfully used to correct for BCG as well as for GA, when one suspect of subject's motion or when cardiac events are difficult to detect, more sophisticated approaches are advisable [31].

11 Simultaneous EEG-fMRI: Integrating Spatial and Temporal Resolution 205

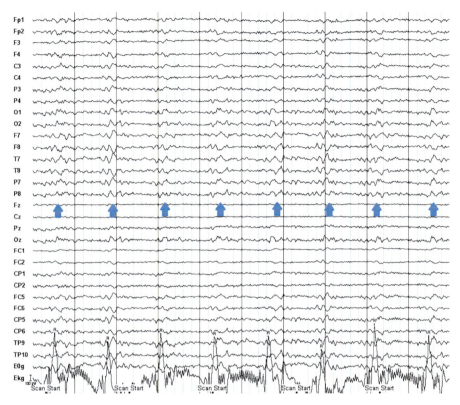

Fig. 11.3 EEG traces contaminated by pulse artifact, also called ballistocardiogram (BCG). Pulse artifacts are visible (see arrows) in most EEG channels and are phase shifted with respect to the R signal in the ECG

Similar to the need of conditioning the EEG signal, fMRI also has preliminary pre-processing steps, which usually involve slice scan time correction, spatial and temporal filtering, and motion correction. Following this process, statistics are used to characterize the BOLD signal. The most common approach used to identify voxels with significant BOLD response is the General Linear Model (GLM) [24]. This framework was first developed for PET studies and then extended for fMRI analysis. The aim of the GLM is to explain the variance of the time course, y_i, $i = 1, \ldots, N$ (N is the total number of observations), as a weighted linear combination related to the effects of interest. In general, multiple regression procedures will estimate a linear equation of the form:

$$y_i = b_1 X_{i_1} + b_2 X_{i_2} + \cdots + b_k X_{i_k} + \varepsilon_i,$$

where X is the predictor models for each k effect, b are the weight parameters and ε_i is the residual error. The regression model in matrix notation can be expressed as:

$$Y = XB + \varepsilon.$$

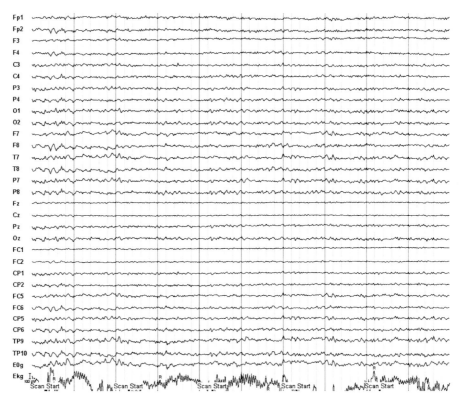

Fig. 11.4 Same segment of EEG traces shown in Fig. 11.3, after the pulse artifact and gradient corrections

Now, Y is the vector with voxels values, X is known as the design matrix, B is the vector of parameters, with a column vector of 1 (for the intercept) plus k unknown regression coefficients, and ε is the vector of error terms. The goal of the GLM is to minimize the sum of squared residuals and find the regression coefficients that satisfy this criterion.

The design matrix is constructed based on models that try to resemble the effects of interest. It is usually obtained by convolving a simplified representation of the event with a canonical hemodynamic response function (HRF). For example, in the case of block paradigms it can be modeled as a boxcar function with the same duration of the external stimulus.

The results of the GLM analysis are based on t or F statistics, depending on whether one parameter or several parameters are being tested at each voxel. Usually, statistic values are compared against the null hypothesis, which is the distribution of the values if there was no effect. Considering that statistics over the noise distribution is known, only the false-positive rate, or type I error, can be controlled. In this respect, a number of methods have been created to address the issue of multiple comparisons in functional neuroimaging, which either place limits on the family-wise error rate (FWER) or the false discovery rate (FDR). Bonferroni correction and Gaussian Random Field Theory (GRFT) [79] are used to control FWER. Bonferroni correction is too conservative for most functional neuroimaging, since it has some degree of spatial dependence between voxels. GRFT can overcome this dependency by using a spatial smoothing filter. Using these methods with a cutoff value of 0.05 means a 5% chance of false positives across the entire set of tests. On the other hand, False Discovery Rate is a less conservative alternative when compared to family-wise statistic [11]. It uses a threshold that allows a given portion of significant voxels to correspond to false detections. In other words, using a cutoff value of 0.05 means that on average 5% of the detected voxels can be false positives.

11.6 Clinical Application: Epilepsy

Approximately 9% of all patients with epilepsy suffer from refractory focal epilepsy syndromes [65] and about half of such patients are potential candidates for surgical epilepsy treatment. Depending on the specificities of the epileptic syndrome and on the efficiency to define and remove the epileptogenic zone, surgical treatment for epilepsy has been an important and effective way to control seizures. Large epilepsy centers have reported an average rate of 60% of patients who underwent surgery and that were free of seizures [21].

The main goal of ablative epilepsy surgery is the resection of the epileptogenic zone (EZ) with preservation of eloquent cortex. The EZ, though, is a theoretical concept that describes the area of abnormal cortex necessary for the generation of seizures, whose complete resection or complete disconnection results in the elimination of seizures. Thus, if the patient is free of seizures after surgery, one may conclude that EZ was successfully included in the resection. There is no currently available diagnostic modality that can be applied to measure the entire EZ directly. However, the location and extend of the area can be elucidated indirectly by using information gained from clinical semiology, EEG (ictal and interictal, and in some cases intracranial), video-EEG, psychological tests and neuroimages. These diagnostic tools define different cortical zones: symptomatogenic zone, irritative zone, zone of early ictal, functional deficit zone and epileptogenic lesion. Each one provides information about the location and extension of the EZ, depending on the sensitivity and specificity of the diagnosis method applied.

These techniques are invasive when, for example, depth electrodes are needed or noninvasive, with minimal physical interference to the patient. Surgical epilepsy treatment employs a variety of non-invasive imaging procedures including structural imaging (MRI), functional imaging (fMRI, PET and Single Photon Emission Computed Tomography—SPECT), and neurophysiology (EEG and MEG). Usually, combined information is obtained by coincidence of specific signals across different modalities.

EEG has been used in the context of epileptic disorders since its discovery. These studies are critical in the diagnosis of epilepsy, classification of epileptic disorders and in locating the sources of epileptic activity. However, as mentioned before, the

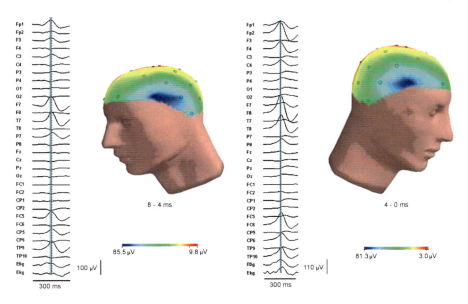

Fig. 11.5 Scalp EEG of an epileptic patient. Mean interictal epileptic events located bilaterally, in channels located in left and right temporal lobe. 3D map projection of the EEG signal amplitude in a 4 ms window centered at the main peak of the spikes

location of EEG sources is limited by its low spatial resolution. Thus, to improve the diagnostic characterization and localization of EZ, EEG is sometimes associated with behavioral data, generated by a video, setting up a video-EEG facility. In defining the irritative zone and the zone that triggers the seizures, video-EEG monitoring remains the gold standard.

In traditional EEG analysis of epileptic events, the average of all spikes and its respective 3D map projection onto the EEG reference system help giving a broad topographical localization of the spikes (Fig. 11.5).

Neuroimaging studies, such as MRI, have contributed greatly to the definition of different epileptic zones. MRI allows the identification of structural changes with excellent spatial resolution, identifying atrophic lesions, dysplasic or neoplastic [5, 14]. In recent years, fMRI has played an important role in the evaluation of cognitive function and preoperative evaluation of patients with epilepsy [6, 71].

Despite all diagnostic tools available, a significant number of patients are not submitted to surgical treatment as a consequence of the difficulty to correctly delineate the EZ. Therefore, the development of new strategies should help to improve epilepsy surgical decision. Accordingly, simultaneous EEG-fMRI has been emerging as an interesting alternative diagnostic toll for epilepsy. In one hand, EEG provides the high temporal resolution necessary to detect epileptiform discharges, on the other fMRI comes with a higher spatial precision to locate the irritative zone (IZ). Therefore, in order to help resembling the EZ, the primary aim of EEG-fMRI is the identification of interictal epileptic discharges (IED) that are transient events observed in the EEG of epileptic patients. These are divided according to its morphology, as follows [19]: spikes (fast transients signal lasting less than 70 ms with an electronegative peak); sharp waves (transient signal lasting about 70–120 ms, with a sharp peak), spike-and-slow-wave complexes (consisting of a spike followed by a slow wave), and polyspike-wave (PSW) complexes (also called multiple-spike-and-slow-wave complexes, same as spike-and-slow-wave complex, but with 2 or more spikes associated with one or more slow waves). IED may occur in isolation or in brief bursts.

Focal discharges usually manifest isolated as sharp waves or spikes, and are often associated with slow waves generally related only to one hemisphere. Otherwise, generalized discharges, usually composed by spike-wave or polyspike-wave complexes, involving both hemispheres.

Early studies of combined EEG-fMRI in epilepsy were typically achieved by triggering the onset of fMRI acquisition to each spike detected in the EEG [40, 69, 77]. Using this strategy, EEG-triggered fMRI is not influenced by gradient artifact, preserving the EEG time series reasonably clean between scanning periods. Nonetheless, this approach has some limitations [69]: only a limited number of images can be acquired, which compromises the construction of a baseline period; the EEG signal during MR acquisition is lost and, therefore, it is not possible to ascertain the absence of other IED during that period; it is necessary to have an online EEG monitoring by a trained neurophysiologist during the entire acquisition session.

The promising potential demonstrated by these first results boosted the development both of acquisition hardware and artifact reduction algorithms [4, 46]. The main focus was to allow continuous recording of EEG-fMRI in a way as to filter, either online or offline, the artifacts added to EEG traces by MRI acquisition, as well as to have the images free of artifact from the EEG system.

As a consequence of the low specificity of automatic detection strategies, IED are usually marked and classified upon visual inspection by a trained neurophysiologist. The number and location of the IED is random and cannot be controlled, limiting patient selection and scanning strategies.

Currently, the localization of IED correlated with a BOLD signal is most often based on the GLM [24]. In such, the timings of the IED are modeled as delta functions and normal EEG periods are considered for baseline states. The design matrix is then built by convolving each of the selected delta function with a canonical model of the HRF. However, at present the precise form of BOLD response to epileptic events is not known. Some studies suggest that these HRF are primarily canonical [47] while other shows non-canonical responses [52]. A flexible approach uses a set of HRF, allowing greater shape variations [30] that can improve the sensitivity to detect non-canonical BOLD responses. However, those have been related to an increased detection of false-positives [75], and have, therefore, been associated to the use of random field theory [80] or the false discovery rate methods [11].

Usually, studies have selected only patients with frequent IED observed on the routine EEG. Despite of this, there is a great chance of not detecting any IED during the EEG-fMRI session. Indeed, Al-Asmi et al. [2] and Salek-Haddadi et al. [66]

observed IEDs only in about 50% of the selected patients and Lemieux et al. [47] observed in 62%. Moreover, neurophysiologists tend to exclude discharges that are not clearly epileptiform, limiting, therefore, the number of IED marked [22]. In order to improve IED detection, a number of alternative computational methods have emerged. A promising one is based on Independent Component Analysis (ICA), which includes information of amplitude, duration and topography of IED [39]. Moreover, sub-threshold epileptic activities can be considered and may increase the sensitivity and statistical power of EEG-fMRI analysis. Using this approach, Jann and colleagues [36] showed an improved sensitivity of BOLD responses detection, from 50% to 80%, when compared to IED manually identified.

Another important issue has been the low sensitivity of EEG-fMRI. Only about 40–70% of studies show a significant spike-related BOLD response. For instance, Salek-Haddadi and colleagues [66] reported that spike-related BOLD changes were detected in 68% of patients from those that discharges were observed during acquisition. This is comparable with a study by Krakow et al. [41], in which only 58% of patients had BOLD response related to the selected IED. Additionally, studies using a canonical HRF observed spikes-associated activation in about 40% of the patients studied [47].

A possible cause for the observed low sensitivity is the critical dependency of the GLM on the HRF model. In fact, experiments have shown that the HRF differs within and between subjects as seen in normal controls [1] and patients [53]. Therefore, particularly in EEG-fMRI analysis, a confounding factor might be related to HRF variability in epileptic tissue. In fact, some studies have brought about evidence that spike-related HRF morphology are more variable than those from normal brain tissue [10, 35, 37, 52].

Another issue relates to the total number of spikes for which one can obtain a consistent BOLD response. The amplitude of the HRF significantly decreases as the number of spikes increases, indicating that spike-related BOLD changes may not give support to the linearity assumption needed by most statistical detection methods. On the other hand, if this number is to small the statistical power is decreased [29, 35, 66].

Therefore, when dealing with pathological events such as epileptic spikes, classical statistical approaches using traditional HRF may not be best choice. Consequently, a series of alternative approaches have been developed in order to better estimate the HRF. One possibility to improve detection consists of using a series of gamma functions, with peaks at different latencies (i.e., 3, 5, 7 and 9 seconds). Using this approach, an increased detection of 17.5% was obtained when compared to standard methods [7]. A similar approach increased the detection power in 90% [38], however, it is important to note that such improvement might come at a cost of less specificity. Another strategy is to use a patient-specific HRF, which also improves EEG-fMRI results [37]. Indeed, Masterton et al. [52] used the group-average BOLD response from a homogeneous cohort of patients and got a significant increase in specificity and sensitivity, from 44% to 89%, when compared to the canonical HRF. Another method uses half the total number of spikes to estimate a voxel-specific HRF and the second half is used to detect the correspondent BOLD responses [50].

11 Simultaneous EEG-fMRI: Integrating Spatial and Temporal Resolution

Additionally, there have been created a number of strategies that does not rely on a specific HRF model of the BOLD response. Interactive temporal clustering analysis is one of these techniques and is not based on the timing of the events, but indeed in the temporal similarity of a spatially confined set of voxels [25]. Another promising technique is based on Independent Component Analysis, in which the dataset is spatially decomposed into independent components to identify epileptic activity, independent of the EEG [64].

In the last years, alternative methods based on measurements of information theory have been used for analysis conventional fMRI [18, 73]. Among different metrics to measure information in fMRI time series, a generalized Kullback–Leiber (KL) distance implementation was the one that gave the best results [13]. The main advantage of this strategy is the fact that it does not assume a specific shape for the HRF but instead only considers the general structure of the signal. The KL method has a great deal of flexibility and variability in the type of HRF that can be detected, which would be of great interest to EEG-fMRI analysis.

The aim of the KL is to measure a distance between the likelihood functions p_1 and p_2 of discrete random variables X_1 and X_2, respectively. X_1 are composed by the segments of the time courses, defined by a fixed window, in which is expected the BOLD response. On the other hand X_2 include the remaining intervals, to be mainly composed by baseline periods. The generalized Kullback–Leibler distance has the form:

$$D_q = \frac{k}{(1-q)} \sum_{j=1}^{L} p_{1j}^q \left(p_{1j}^{1-q} - p_{2j}^{1-q} \right),$$

k is a positive constant and its value depends on the unit used. For instance, if $k = 1/\ln 2$ the entropy is measured in bits. The q value (Tsallis parameter) is a real number, and when $q \to 1$, $D_q \to D$, and Shannon entropy is recovered.

The goal of the KL is to calculate a distance for each spike-related BOLD response from baseline, and to generate a map that corresponds to the average KL distance ($\bar{D}_q$). As the distribution of the coefficients under the null hypothesis is unknown, the statistical significance of this variable is determined by directly using a phase permutation test [43]. An example of KL statistical maps for a patient with focal bilateral, right and left temporal lobe epilepsy, are shown in Figs. 11.6 (left hemisphere) and 11.7 (right hemisphere). For comparison purposes, maps obtained from the same patients using the GLM were also obtained (Figs. 11.6 and 11.7). The selected threshold was corrected for multiple comparisons, which was based on the cluster size [23], generated by Monte Carlo simulations.

The map shown in red corresponds to areas detected by the KL and not by the GLM. The reverse case is shown in blue. Note that although the areas extracted by KL and GLM seem to mostly overlap (brown areas represent overlap between GLM and KL), some regions are better identified by KL and others by GLM. Nonetheless, the maps obtained by both methods confirm the expectation that most relevant discriminatory information is located in the temporal lobe.

The main difference between KL and GLM is that, while GLM use an a priori canonical model, KL shares the advantage of not requiring any HRF model for real

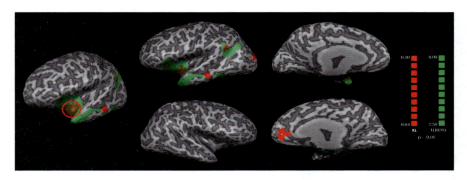

Fig. 11.6 fMRI statistical maps of spikes detected in the EEG from the left temporal lobe of a patient with bilateral focal epilepsy. *Red circle* shows the localization on the left hemisphere determined by both methods (KL and GLM). In *green GLM*—BOLD increase; in *red* the maps from KL, in *brown* is the overlay between GLM and KL ($p < 0.05$ corrected)

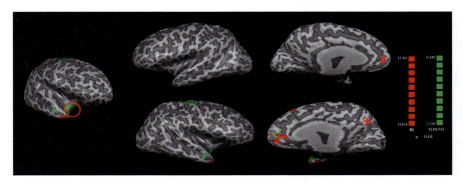

Fig. 11.7 fMRI statistical maps of the spikes detected in the EEG from the right temporal lobe of a patient with bilateral focal epilepsy. *Red circle* shows the localization on right hemisphere determined by both methods (KL and GLM). In *green GLM*—BOLD increase; in *red* the maps from KL, in *brown* is the overlay between GLM and KL ($p < 0.05$ corrected)

data analysis. As a consequence, it is expected that GLM present a greater detection power than KL when BOLD responses match the chosen HRF model. Conversely, when BOLD responses diverge from canonical, KL appears to be an interesting alterative.

The concordance between EEG and fMRI has been generally reported by comparing fMRI results with scalp topography of the discharges (Fig. 11.5). Additionally, a comparison between fMRI and EEG source localization is becoming more widespread. It has been shown that EEG source localization using distributed source methods is probably the most appropriate to recover spike generators with their spatial extent [8]. Using this approach, Grova et al. [32] showed that, for most patients, part of the BOLD responses were highly concordant with EEG sources calculated along the duration of the spike. Again, the limited sensitivity of scalp EEG is an important factor to be considered, as the signal detected is attenuated and distorted by different head tissue such as skull and scalp. For instance, it has been shown that scalp EEG detects only about 9% of the discharges identified using intracranial electrodes [56].

Another important limitation on multimodal EEG-fMRI integration is imposed by physiology. A mismatch between the detected EEG and fMRI sources are sometimes observed since EEG measurements direct results from neuronal electrical activity, while fMRI is based on the vascular response that follows neuronal activity. In addition to pre- and post-synaptic activity, other physiological processes that require energetic supply, such neurotransmitter synthesis [61] and glial cell metabolism [45] may cause BOLD changes without EEG correlates. Therefore, in general, one cannot expect a perfect correspondence between the information provided by EEG and fMRI, but instead understand that these techniques provide important complementary information [28].

11.7 Conclusion

EEG-fMRI is attracting continuous attention from the neuroscience community and has been emerging as promising technique for basic and clinical inspection. For instance, as demonstrated herein, this multimodal approach can reveal whole-brain maps of haemodynamic changes that are correlated with pathological EEG patterns, such as those from IED. Moreover, it has been successfully applied to inspect different aspects of human brain oscillations [55], which includes investigating human sleep [16, 78]. There are still, however, a number of technical limitations, which are to be addresses in the near future. In particular the limited sensitivity of fMRI, which sometimes results from the mismatch between actual BOLD responses and the canonical model used. The application of alternative fMRI analysis techniques as well as the improvement of EEG and fMRI recording techniques may improve sensitivity and provide a better understanding of the relationship between EEG and BOLD signals. Thus, the combination between technological development and careful interpretation of the complementary spatiotemporal patterns derived from the EEG and fMRI signals will certainly lead to new insights of human brain function.

References

1. Aguirre GK, Zarahn E, D'Esposito M (1998) The variability of human, BOLD hemodynamic responses. NeuroImage 8:360–369
2. Al-Asmi A, Benar CG, Gross DW, Khani YA, Andermann F, Pike B, Dubeau F, Gotman J (2003) fMRI activation in continuous and spike-triggered EEG-fMRI studies of epileptic spikes. Epilepsia 44:1328–1339
3. Allen PJ, Polizzi G, Krakow K, Fish DR, Lemieux L (1998) Identification of EEG events in the MR scanner: the problem of pulse artifact and a method for its subtraction. NeuroImage 8:229–239
4. Allen PJ, Josephs O, Turner R (2000) A method for removing imaging artifact from continuous EEG recorded during functional MRI. NeuroImage 12:230–239

5. Antel SB, Collins DL, Bernasconi N, Andermann F, Shinghal R, Kearney RE, Arnold DL, Bernasconi A (2003) Automated detection of focal cortical dysplasia lesions using computational models of their MRI characteristics and texture analysis. NeuroImage 19:1748–1759
6. Araujo D, de Araujo DB, Pontes-Neto OM, Escorsi-Rosset S, Simao GN, Wichert-Ana L, Velasco TR, Sakamoto AC, Leite JP, Santos AC (2006) Language and motor FMRI activation in polymicrogyric cortex. Epilepsia 47:589–592
7. Bagshaw AP, Aghakhani Y, Benar CG, Kobayashi E, Hawco C, Dubeau F, Pike GB, Gotman J (2004) EEG-fMRI of focal epileptic spikes: analysis with multiple haemodynamic functions and comparison with gadolinium-enhanced MR angiograms. Hum Brain Mapp 22:179–192
8. Bagshaw AP, Kobayashi E, Dubeau F, Pike GB, Gotman J (2006) Correspondence between EEG-fMRI and EEG dipole localisation of interictal discharges in focal epilepsy. NeuroImage 30:417–425
9. Benar C, Aghakhani Y, Wang Y, Izenberg A, Al-Asmi A, Dubeau F, Gotman J (2003) Quality of EEG in simultaneous EEG-fMRI for epilepsy. Clin Neurophysiol 114:569–580
10. Benar CG, Gross DW, Wang Y, Petre V, Pike B, Dubeau F, Gotman J (2002) The BOLD response to interictal epileptiform discharges. NeuroImage 17:1182–1192
11. Benjamini Y, Hochberg Y (1995) Controlling the false discovery rate—a practical and powerful approach to multiple testing. J R Stat Soc B, Methodol 57:289–300
12. Bonmassar G, Anami K, Ives J, Belliveau JW (1999) Visual evoked potential (VEP) measured by simultaneous 64-channel EEG and 3T fMRI. Neuroreport 10:1893–1897
13. Cabella BCT, Sturzbecher MJ, de Araujo DB, Neves UPC (2009) Generalized relative entropy in functional magnetic resonance imaging. Physica a, Stat Mech Appl 388:41–50
14. Cendes F, Caramanos Z, Andermann F, Dubeau F, Arnold DL (1997) Proton magnetic resonance spectroscopic imaging and magnetic resonance imaging volumetry in the lateralization of temporal lobe epilepsy: a series of 100 patients. Ann Neurol 42:737–746
15. Christmann C, Koeppe C, Braus DF, Ruf M, Flor H (2007) A simultaneous EEG-fMRI study of painful electric stimulation. NeuroImage 34:1428–1437
16. Czisch M (2008) The functional significance of K-complexes: new insights by fMRI. J Sleep Res 17:6
17. Daunizeau J, Grova C, Marrelec G, Mattout J, Jbabdi S, Pelegrini-Issac M, Lina JM, Benali H (2007) Symmetrical event-related EEG/fMRI information fusion in a variational Bayesian framework. NeuroImage 36:69–87
18. de Araujo DB, Tedeschi W, Santos AC, Elias J Jr, Neves UP, Baffa O (2003) Shannon entropy applied to the analysis of event-related fMRI time series. NeuroImage 20:311–317
19. de Curtis M, Avanzini G (2001) Interictal spikes in focal epileptogenesis. Prog Neurobiol 63:541–567
20. de Munck JC, Goncalves SI, Huijboom L, Kuijer JP, Pouwels PJ, Heethaar RM, Lopes da Silva FH (2007) The hemodynamic response of the alpha rhythm: an EEG/fMRI study. NeuroImage 35:1142–1151
21. Engel J Jr (1993) Clinical neurophysiology, neuroimaging, and the surgical treatment of epilepsy. Curr Opin Neurol Neurosurg 6:240–249
22. Flanagan D, Abbott DF, Jackson GD (2009) How wrong can we be? The effect of inaccurate mark-up of EEG/fMRI studies in epilepsy. Clin Neurophysiol 120:1637–1647
23. Forman SD, Cohen JD, Fitzgerald M, Eddy WF, Mintun MA, Noll DC (1995) Improved assessment of significant activation in functional magnetic resonance imaging (fMRI): use of a cluster-size threshold. Magn Reson Med 33:636–647
24. Friston KJ, Holmes AP, Poline JB, Grasby PJ, Williams SC, Frackowiak RS, Turner R (1995) Analysis of fMRI time-series revisited. NeuroImage 2:45–53
25. Gao JH, Yee SH (2003) Iterative temporal clustering analysis for the detection of multiple response peaks in fMRI. J Magn Reson Imaging 21:51–53
26. Goense JB, Logothetis NK (2008) Neurophysiology of the BOLD fMRI signal in awake monkeys. Curr Biol 18:631–640
27. Goldman RI, Stern JM, Engel J Jr, Cohen MS (2002) Simultaneous EEG and fMRI of the alpha rhythm. Neuroreport 13:2487–2492

28. Gotman J (2008) Epileptic networks studied with EEG-fMRI. Epilepsia 49(Suppl 3):42–51
29. Gotman J, Benar CG, Dubeau F (2004) Combining EEG and FMRI in epilepsy: methodological challenges and clinical results. J Clin Neurophysiol 21:229–240
30. Goutte C, Nielsen FA, Hansen LK (2000) Modeling the haemodynamic response in fMRI using smooth FIR filters. IEEE Trans Med Imaging 19:1188–1201
31. Grouiller F, Vercueil L, Krainik A, Segebarth C, Kahane P, David O (2007) A comparative study of different artefact removal algorithms for EEG signals acquired during functional MRI. NeuroImage 38:124–137
32. Grova C, Daunizeau J, Kobayashi E, Bagshaw AP, Lina JM, Dubeau F, Gotman J (2008) Concordance between distributed EEG source localization and simultaneous EEG-fMRI studies of epileptic spikes. NeuroImage 39:755–774
33. Heeger DJ, Huk AC, Geisler WS, Albrecht DG (2000) Spikes versus BOLD: what does neuroimaging tell us about neuronal activity? Nat Neurosci 3:631–633
34. Ives JR, Warach S, Schmitt F, Edelman RR, Schomer DL (1993) Monitoring the patient's EEG during echo planar MRI. Electroencephalogr Clin Neurophysiol 87:417–420
35. Jacobs J, Hawco C, Kobayashi E, Boor R, LeVan P, Stephani U, Siniatchkin M, Gotman J (2008) Variability of the hemodynamic response as a function of age and frequency of epileptic discharge in children with epilepsy. NeuroImage 40:601–614
36. Jann K, Wiest R, Hauf M, Meyer K, Boesch C, Mathis J, Schroth G, Dierks T, Koenig T (2008) BOLD correlates of continuously fluctuating epileptic activity isolated by independent component analysis. NeuroImage 42:635–648
37. Kang JK, Benar C, Al-Asmi A, Khani YA, Pike GB, Dubeau F, Gotman J (2003) Using patient-specific hemodynamic response functions in combined EEG-fMRI studies in epilepsy. NeuroImage 20:1162–1170
38. Kobayashi E, Bagshaw AP, Grova C, Dubeau F, Gotman J (2006) Negative BOLD responses to epileptic spikes. Hum Brain Mapp 27:488–497
39. Kobayashi K, James CJ, Nakahori T, Akiyama T, Gotman J (1999) Isolation of epileptiform discharges from unaveraged EEG by independent component analysis. Clin Neurophysiol 110:1755–1763
40. Krakow K, Woermann FG, Symms MR, Allen PJ, Lemieux L, Barker GJ, Duncan JS, Fish DR (1999) EEG-triggered functional MRI of interictal epileptiform activity in patients with partial seizures. Brain 122(Pt 9):1679–1688
41. Krakow K, Lemieux L, Messina D, Scott CA, Symms MR, Duncan JS, Fish DR (2001) Spatio-temporal imaging of focal interictal epileptiform activity using EEG-triggered functional MRI. Epileptic Disord 3:67–74
42. Kwong KK, Belliveau JW, Chesler DA, Goldberg IE, Weisskoff RM, Poncelet BP, Kennedy DN, Hoppel BE, Cohen MS, Turner R et al (1992) Dynamic magnetic resonance imaging of human brain activity during primary sensory stimulation. Proc Natl Acad Sci USA 89:5675–5679
43. Lahiri SN (2010) Resampling methods for dependent data. Springer, New York
44. Laufs H, Krakow K, Sterzer P, Eger E, Beyerle A, Salek-Haddadi A, Kleinschmidt A (2003) Electroencephalographic signatures of attentional and cognitive default modes in spontaneous brain activity fluctuations at rest. Proc Natl Acad Sci USA 100:11053–11058
45. Lauritzen M (2005) Reading vascular changes in brain imaging: is dendritic calcium the key? Nat Rev Neurosci 6:77–85
46. Lemieux L, Allen PJ, Franconi F, Symms MR, Fish DR (1997) Recording of EEG during fMRI experiments: patient safety. Magn Reson Med 38:943–952
47. Lemieux L, Laufs H, Carmichael D, Paul JS, Walker MC, Duncan JS (2008) Noncanonical spike-related BOLD responses in focal epilepsy. Hum Brain Mapp 29:329–345
48. Logothetis NK (2002) The neural basis of the blood-oxygen-level-dependent functional magnetic resonance imaging signal. Philos Trans R Soc Lond B, Biol Sci 357:1003–1037
49. Logothetis NK, Pauls J, Augath M, Trinath T, Oeltermann A (2001) Neurophysiological investigation of the basis of the fMRI signal. Nature 412:150–157

50. Lu Y, Grova C, Kobayashi E, Dubeau F, Gotman J (2007) Using voxel-specific hemodynamic response function in EEG-fMRI data analysis: An estimation and detection model. NeuroImage 34:195–203
51. Makiranta MJ, Ruohonen J, Suominen K, Sonkajarvi E, Salomaki T, Kiviniemi V, Seppanen T, Alahuhta S, Jantti V, Tervonen O (2004) BOLD-contrast functional MRI signal changes related to intermittent rhythmic delta activity in EEG during voluntary hyperventilation-simultaneous EEG and fMRI study. NeuroImage 22:222–231
52. Masterton RA, Harvey AS, Archer JS, Lillywhite LM, Abbott DF, Scheffer IE, Jackson GD (2010) Focal epileptiform spikes do not show a canonical BOLD response in patients with benign rolandic epilepsy (BECTS). NeuroImage 51:252–260
53. Mazzetto-Betti KC, Pontes-Neto OM, Leoni RF, Santos AD, Silva AC, Leite JP, Araujo DB (2010) BOLD response stability is altered patients with chronic ischemic stroke. Stroke 41:E370
54. Mukamel R, Gelbard H, Arieli A, Hasson U, Fried I, Malach R (2005) Coupling between neuronal firing, field potentials, and FMRI in human auditory cortex. Science 309:951–954
55. Musso F, Brinkmeyer J, Mobascher A, Warbrick T, Winterer G (2010) Spontaneous brain activity and EEG microstates. A novel EEG/fMRI analysis approach to explore resting-state networks. NeuroImage 52:1149–1161
56. Nayak D, Valentin A, Alarcon G, Garcia Seoane JJ, Brunnhuber F, Juler J, Polkey CE, Binnie CD (2004) Characteristics of scalp electrical fields associated with deep medial temporal epileptiform discharges. Clin Neurophysiol 115:1423–1435
57. Niazy RK, Beckmann CF, Iannetti GD, Brady JM, Smith SM (2005) Removal of FMRI environment artifacts from EEG data using optimal basis sets. NeuroImage 28:720–737
58. Ogawa S, Lee TM, Kay AR, Tank DW (1990) Brain magnetic resonance imaging with contrast dependent on blood oxygenation. Proc Natl Acad Sci USA 87:9868–9872
59. Ogawa S, Tank DW, Menon R, Ellermann JM, Kim SG, Merkle H, Ugurbil K (1992) Intrinsic signal changes accompanying sensory stimulation: functional brain mapping with magnetic resonance imaging. Proc Natl Acad Sci USA 89:5951–5955
60. Ou W, Nummenmaa A, Golland P, Hamalainen MS (2009) Multimodal functional imaging using fMRI-informed regional EEG/MEG source estimation. Conf Proc IEEE Eng Med Biol Soc 2009:1926–1929
61. Patel AB, de Graaf RA, Mason GF, Kanamatsu T, Rothman DL, Shulman RG, Behar KL (2004) Glutamatergic neurotransmission and neuronal glucose oxidation are coupled during intense neuronal activation. J Cereb Blood Flow Metab 24:972–985
62. Pauling L, Coryell CD (1936) The magnetic properties and structure of hemoglobin, oxyhemoglobin and carbonmonoxyhemoglobin. Proc Natl Acad Sci USA 22:210–216
63. Rauch A, Rainer G, Logothetis NK (2008) The effect of a serotonin-induced dissociation between spiking and perisynaptic activity on BOLD functional MRI. Proc Natl Acad Sci USA 105:6759–6764
64. Rodionov R, De Martino F, Laufs H, Carmichael DW, Formisano E, Walker M, Duncan JS, Lemieux L (2007) Independent component analysis of interictal fMRI in focal epilepsy: comparison with general linear model-based EEG-correlated fMRI. NeuroImage 38:488–500
65. Rosenow F, Luders H (2001) Presurgical evaluation of epilepsy. Brain 124:1683–1700
66. Salek-Haddadi A, Diehl B, Hamandi K, Merschhemke M, Liston A, Friston K, Duncan JS, Fish DR, Lemieux L (2006) Hemodynamic correlates of epileptiform discharges: an EEG-fMRI study of 63 patients with focal epilepsy. Brain Res 1088:148–166
67. Sanei SACJ (2007) EEG signal processing. Wiley, New York
68. Sato JR, Rondinoni C, Sturzbecher M, de Araujo DB, Amaro E Jr (2010) From EEG to BOLD: brain mapping and estimating transfer functions in simultaneous EEG-fMRI acquisitions. NeuroImage 50:1416–1426
69. Seeck M, Lazeyras F, Michel CM, Blanke O, Gericke CA, Ives J, Delavelle J, Golay X, Haenggeli CA, de Tribolet N, Landis T (1998) Non-invasive epileptic focus localization using EEG-triggered functional MRI and electromagnetic tomography. Electroencephalogr Clin Neurophysiol 106:508–512

70. Smith AJ, Blumenfeld H, Behar KL, Rothman DL, Shulman RG, Hyder F (2002) Cerebral energetics and spiking frequency: the neurophysiological basis of fMRI. Proc Natl Acad Sci USA 99:10765–10770
71. Springer JA, Binder JR, Hammeke TA, Swanson SJ, Frost JA, Bellgowan PS, Brewer CC, Perry HM, Morris GL, Mueller WM (1999) Language dominance in neurologically normal and epilepsy subjects: a functional MRI study. Brain 122(11):2033–2046
72. Srivastava G, Crottaz-Herbette S, Lau KM, Glover GH, Menon V (2005) ICA-based procedures for removing ballistocardiogram artifacts from EEG data acquired in the MRI scanner. NeuroImage 24:50–60
73. Sturzbecher MJ, Tedeschi W, Cabella BC, Baffa O, Neves UP, de Araujo DB (2009) Non-extensive entropy and the extraction of BOLD spatial information in event-related functional MRI. Phys Med Biol 54:161–174
74. Viswanathan A, Freeman RD (2007) Neurometabolic coupling in cerebral cortex reflects synaptic more than spiking activity. Nat Neurosci 10:1308–1312
75. Waites AB, Shaw ME, Briellmann RS, Labate A, Abbott DF, Jackson GD (2005) How reliable are fMRI-EEG studies of epilepsy? A nonparametric approach to analysis validation and optimization. NeuroImage 24:192–199
76. Wan X, Iwata K, Riera J, Ozaki T, Kitamura M, Kawashima R (2006) Artifact reduction for EEG/fMRI recording: nonlinear reduction of ballistocardiogram artifacts. Clin Neurophysiol 117:668–680
77. Warach S, Ives JR, Schlaug G, Patel MR, Darby DG, Thangaraj V, Edelman RR, Schomer DL (1996) EEG-triggered echo-planar functional MRI in epilepsy. Neurology 47:89–93
78. Wehrle R, Kaufmann C, Wetter TC, Holsboer F, Auer DP, Pollmacher T, Czisch M (2007) Functional microstates within human REM sleep: first evidence from fMRI of a thalamocortical network specific for phasic REM periods. Eur J Neurosci 25:863–871
79. Worsley KJ, Evans AC, Marrett S, Neelin P (1992) A three-dimensional statistical analysis for CBF activation studies in human brain. J Cereb Blood Flow Metab 12:900–918
80. Worsley KJ, Marrett S, Neelin P, Vandal AC, Friston KJ, Evans AC (1996) A unified statistical approach for determining significant signals in images of cerebral activation. Hum Brain Mapp 4:58–73

Index

A
Abstract-time, 135, 139, 142
Abstraction
 design, 154
 hardware, 147, 155
 model, 144
 process, 145
 software, 155
 spike(s), 138, 142, 146
 time, 142, 148
Acceleration, 33, 35, 37, 39, 41–47, 49–51, 53, 54
Adaptive gains, 22
Adaptive scheme, 15
Adaptive synchronization, 12
Address-event representation (AER), 138, 142, 156
Algorithms, 182
Amplitude, 80
Anatomical studies, 59
Anti-phase synchronization, 125
ARM968, 147–149
Artificial neural networks, 100
Assembly, 33–35, 39, 44, 45, 50–52, 54
Asymptotic stability, 14
Asynchronous, 135, 136, 138, 139, 141–143, 146, 148, 149, 151, 153, 156
Asynchronous event-driven, 155
Asynchronous interactions, 5
Asynchronous real-time, 155
Average artifact subtraction (AAS), 203
Average firing rate, 6

B
Ballistocardiogram (BCG), 203
Binary classifiers, 191
Binding, 34, 50, 51, 54
Binding problem, 64, 78
BOLD (blood oxygenation level dependent), 201
Boundedness, 14

C
Cell body models, 118
Chaotic dynamics, 5, 159–163, 166–168, 172–174, 176, 177
Chaotic itinerancy, 160
Chip multiprocessor (CMP), 135, 137, 138, 143
Class II neuron, 119
Cluttered scenarios, 58
CMOS process, 120
Cognitive brain states, 2
Cohen–Grossberg–Hopfield model, 35, 37, 41
Coherent synchronization, 3
Collective synchronization, 85
Combinatorial explosion, 57
Competition for coherence, 33, 34, 37, 41, 43–48, 50–52, 54
Complex networks, 75
Complex object, 181
Complex problem, 175–177
Complex problems, 160, 161
Concurrency, 144
Concurrent, 135, 136, 138, 140–142, 145, 146, 149, 154, 155, *see also* parallel simulation, 144
Concurrent language, 144
Configurable, 135, 136, 138, 139, 143, 144, 147, 156, *see also* programmable
Connectivity map, 76
Connectivity patterns, 80
Constrained chaos, 160

Contracting, 12
Contracting input, 25
Contracting systems, 12
Contraction rate, 12
Contraction theory, 9, 16
Coordination of motion, 9
Cortical structure, 78
Coupled P-HNMs, 117, 120
Coupling functions, 3
CPG model, 117
Cycle accurate, 144, 148, 150
Cycle approximate, 148

D
Delay, 113
Delayed feedback responses, 92
Desynchronization, 39, 44
Device level, 144, 147
Diffusive coupling, 26
Droppings, 103
Dynamic environment, 113
Dynamical systems, 3

E
Edge-based strategy, 13
Electric field potential, 201
Electroencephalography (EEG), 6, 200
Environment, 111, 112
Epileptic discharges, 6
Epileptogenic zone (EZ), 207
Event, 139–142, 146, 147
    driven, 136, 139–141, 146, 147, 149, 153
    driven processing, 137, 141
Evidence for top-down feedback, 59
Evolution, 106
Evolutionary, 103
Evolutionary origin, 112
Evolutionary origin of recollection and prediction, 101
Evolved, 113
Evolving neural networks, 100
Evolving the network structure, 15
Excitatory, 92
Excitatory mutual coupling, 123
Excitatory-only coupled neurons, 28
Excitatory-only coupling, 25
External marker, 102, 113
Extrapolation, 113

F
Feedback connections, 76
Feedback inhibition, 4
Feedforward, 76, 100, 103
Field potentials, 25

Firing rate, 183
Fitness, 113
FitzHugh–Nagumo (FN) oscillators, 19, 25
FitzHugh–Nagumo neurons, 10, 19
Freely behaving, 183
Frequency, 80
Function pipeline, 153, 154, 156
Functional magnetic resonance imaging (fMRI), 6, 200

G
Gain adaptation law, 13
Gamma oscillations, 52
General linear model (GLM), 6, 205
Generalized Kullback–Leiber (KL) distance, 211
Generative models, 57
Genetic algorithms, 103
Global synchrony, 85
Gradient artifacts (GA), 202
Grid, 188
Grid computing, 188

H
Hardware, 135, 137, 139, 140, 143, 144, 146–149, 151, 154–156
Hardware description language (HDL), 135, 144, 148, 154
Hardware design environment, 136
Hardware design flow, 155
Hardware design systems, 154
Hardware solutions, 2
Hardware-based neural models, 5
Hebbian memory, 33–36, 38, 39, 44, 45, 49, 51, 54
Hemodynamic response function (HRF), 206
Hierarchical pathways, 78
Hippocampus, 111, 112
Homeostatic plasticity, 59

I
IC chip, 117
Ill-posed problem, 159, 161, 163, 172, 176
In-phase synchronization, 124
Incoherent memory, 137, 139, 140
Incremental reconfiguration, 137, 142
Inhibitory, 92
Inhibitory mutual coupling, 124
Interictal epileptic discharges (IED), 208
Internal state dynamics, 107
Internal state predictability, 110, 111
Internal state trajectory, 101, 108
Izhikevich, 153
Izhikevich (model), 150, 151

Index

K
Kuramoto model, 37–39

L
Large-scale networks, 2
Lateral, 76
Layout pattern, 131
Leaky-integrate-and-fire, 153
Learning, 83
Libraries, 135, 144, 146, 147, 149, 154
Local field potentials, 26
Locomotion rhythms, 117

M
Material interaction, 101
Maximization, 80
Maze navigation, 5
Memory, 100, 106
MEMS (Micro Electro Mechanical Systems) type robot, 117
Model level, 144–147, 155
Model(s) of computation, 136
Modularity, 61
Multi-electrode technique, 5
Multi-layer perceptron (MLP), 190
Multi-neuron, 185

N
Native parallelism, 136, 137, 140
Navigation, 5
Network dynamics, 80
Network topology, 22
Network-on-chip (NoC), 138, 141–143, 148, 149
Neural avalanches, 1
Neural chip multiprocessor, 5
Neural modeling, 1
Neural model(s), 135, 136, 140, 142–144, 146–148, 151, 153–155
Neural objects, 145, 146
Neural oscillators, 9
Neuroevolution, 101, 103
Neuromimetic, 136, 137
Neuromodulators, 112
Neuron dropping, 193
Neuronal ensemble, 189
Neuronal ensembles, 185
Neuronal responses, 183
Neuron(s), 136, 138–143, 145, 146, 149, 150, 153, 155
Nondeterministic, 151, 154, 156

O
Object coding, 6
Object recognition, 78
Objective function, 79
Olfaction, 58, 112
Olfactory system, 112
Origin of the EEG signal, 200
Origin of the fMRI signal, 201
Oscillation, 77
Oscillatory, 36–38, 41, 54
Oscillatory activity, 1

P
Parallel, 136, 138, 140, 144, 155, 156, *see also* concurrent
Partial contraction, 16
Partial synchrony, 9
Partially contracting, 16
Pattern recognition, 33, 34, 36, 37, 39
Pattern-frequency-correspondence, 44, 49
Patterned Coherence Models, 33–35, 37–39, 41, 45, 52–54
Pearson's correlation, 181
Perceptual, 2
Perceptual binding, 2
Phase, 80
Phase information, 84
Phase plane, 119
Phase segmentation, 84
Phase-dependent Hebbian learning rule, 82
Phase-locking, 3
Phasor notation, 79
Pheromones, 103
Poisson distribution, 185
Pole balancing, 101, 106, 107
Precise timing, 1
Predictability, 107, 108
Prediction, 4, 100, 112, 113
Predictive, 106, 111
Presynaptic inhibition, 159, 161, 170, 172–174, 177
Programmable, 135, 136, 140, 142, 143, 146, 153, *see also* configurable
Pulse-type hardware neuron models (P-HNMs), 117
PyNN, 155

Q
QUAD assumption, 11
Quadruped patterns, 127
Quasi-layered, 164, 165, 176
Quasi-layered recurrent neural network, 159
Quorum sensing, 26

R
Rapid-eye-movement sleep (REM), 181
Real-time, 135, 136, 140, 141, 146, 147, 150, 154

Real-time simulations, 2
Recollection, 4, 100, 113
Reconfigurable, 154, 155
Recurrent, 4, 100, 103, 106
Recurrent neural network, 159, 160, 163, 176
Regulatory feedback, 59
Relationship between EEG and fMRI, 201
Relationship information, 84
Robot, 5, 159–163, 168–174, 176, 177
Robustness, 18

S

Scalability, 61, 135, 143
Scalable, 139, 154
Segmentation, 77
Segmentation accuracy, 77, 84
Segmenting, 58
Sensitivity of EEG-fMRI, 210
Separation, 77
Separation accuracy, 84
Signal degradation, 77
Simple rule, 159, 160, 176, 177
Simultaneous EEG-fMRI, 200
Slow-wave sleep (SWS), 181
Sparse representation, 79
Spatial memory, 111
Speedup, 193
Spikes, 181
Spiking, 136, 140, 146, 149, 153
SpiNNaker, 135–149, 151, 153–156
SpiNNaker spiking, 150
Steady state, 83
Steady state equilibrium, 61
Stepping through iterations, 63
Structure-function questions, 92
Superposition, 83
Superposition catastrophe, 57, 67
Surgical epilepsy treatment, 207
Synapse model(s), 140, 141
Synapse(s), 138, 140, 150, 153, 155
Synaptic models, 118
Synaptic objects, 145

Synchronization, 3, 9, 10, 33, 37–39, 41, 42, 47, 50, 51, 53, 54, 77, 83
System level, 144, 146, 147, 155
SystemC, 144–146, 148, 150

T

Template, 135, 144–147
Temporal coding, 33–36, 38–40, 42, 45, 51, 53
Temporal resolution, 6
Testing distribution, 58
Theoretical frameworks, 2
Time, 100, 101
   dynamics, 135, 136, 143
   model, 135, 136, 139, 142, 143, 145–147, 150, 155
   notion of, 154
Timing, 148
Top-down feedback, 59
Topology, 3, 90
Training distribution, 58
Transaction Level Model (TLM), 148
Two-dimensional maze, 159–162, 172–174

U

Unsupervised, 77
User-configurable, 155

V

Vector field, 83
Verification, 148, 149, 154
Vertex-based strategy, 13
Virtual interconnect, 142
Virtually local, 140
Visual cortex, 76
Visual object representation, 77

W

Winner-take-all, 83

Z

Zombies, 112